HORSE BREEDING
New edition

Peter Rossdale

David & Charles

British Library Cataloguing in Publication Data

Rossdale, Peter
   Horse breeding.
   I. Title
   636.1

   ISBN 0-7153-9975-6

First published 1981
Second impression 1983
Revised and extended edition 1992

Typeset by XL Publishing Services Nairn Scotland
and printed in Great Britain
by Redwood Press Ltd Melksham Wilts
for David & Charles
Brunel House Newton Abbot Devon

CONTENTS

ACKNOWLEDGEMENTS

As with any book of this nature, it was necessary for me to obtain assistance from a large number of colleagues and friends. It is not possible to name them all, but they include the management and staff of the studfarms for which I am responsible, in a veterinary capacity. Also, I received much help from members of my practice, particularly Jenny Ousey, Andrew McGladdery and Sidney Ricketts. I would like to thank Jan Wade of R. & W. Publications (Newmarket) Ltd, for her editorial assistance, Rachel Leeks, Anna Arnold and Enrico Stefanelli, a veterinary student from Italy at the TBA Equine Fertility Unit in Newmarket. I would also like to thank Professor O. J. Ginther and Dr W. R. Allen of TBA Equine Fertility Unit for supplying the figures in chapter 14. Finally I must thank Sue Hall, of David & Charles, for her patience and perseverance.

PREFACE

I have written this book for the horse breeder and for those who care for brood mares, stallions or young horses. My aim is to inform the reader, as far as possible in non-technical language, of the biological functions on which equine reproduction is based. It is important for people at all levels of responsibility—owners, stud managers and personnel—to be knowledgeable about the subject, so that they can understand the various happenings they encounter in the practical art of horse breeding, and thereby through a common understanding, be equipped to deal with problems and to communicate with one another on equal terms. It is surprising how often owners cannot comprehend why, for example, their mares are not being mated early in the breeding season, and they frequently blame those on the studfarm for their failure.

The truth about breeding horses is that there are many ways to the same end, and to a great extent people who breed horses have to do so under widely varying circumstances. Some breeders are rich and some are poor, while there are those who are professional and others who are amateurs; some have one mare while others possess many. But all horse breeders have a number of aspirations in common: we want to breed horses in an efficient and proper manner, to succeed, and to maintain the soundness, health and well-being of our horses. It is the duty of veterinarians to explain and it is the purpose of this book to deal with the subject of horse breeding in general terms and not to preach dogma. The person on the spot—in the foaling box or mating shed—has to make the decision in any particular circumstance, guided by background knowledge and information. However, it is necessary to know the facts before we can hope to make a correct decision, whether this decision is to take action or to be patient, to interfere or to call for veterinary assistance.

I have been impressed, during my years in equine veterinary practice, by the enthusiasm of audiences attending lectures on all aspects of horse breeding. Among all types of people responsible for the care of horses, there is obviously a sincere wish to learn, and to improve their knowledge and understanding of the way in which the horse's body functions relate to the day-to-day experiences encountered in terms of sexual activity or inactivity, pregnancy, foaling and neonatal development.

Vets are trained in science and are therefore in a good position to propound the virtues of a scientific approach to the breeding of horses, but it is essential that we all appreciate the effect of our actions in any given circumstance. For example, the natural way of horses is to live as a herd with one male possessing a harem of mares. The stallion and the members of the harem maintain a close social bond throughout their lives and breed during the spring and summer months, so that their foals are born in late spring the following year. For economic, climatic and other reasons we generally keep the stallion separate from his mares and, further, we re-arrange the harem members to suit our own convenience. However, the habit of segregating mares from the stallion imposes a system of manage-ment which entails the ritual of teasing mares to cause them to show sexual signs. In this way it is possible to discern whether or not the individual is in a state conducive to mating with the male horse. But teasing is an unnatural procedure, and mares may respond unfavourably and fail to show typical signs of oestrus (heat). We shall discuss later the routine veterinary examination of mares, likewise other measures which have become necessary in stud management because of the separation of the male from the female horse.

I have attempted, in writing this book, to relate managerial action to biological events, thus, I hope, enabling breeders to judge for themselves the consequences, advantages and disadvantages of procedures that neces-sarily alter the horse's environment, and thereby its response to its particular surroundings. Throughout the text I maintain an assumption that nature's method of breeding must be our reference point; and that we cannot improve on nature but only co-operate in a manner that is least harmful. To achieve this end we must be aware of natural function; and Part II of the book is devoted to this aspect. Part III is concerned with abnormalities and disease.

A decade has passed since I wrote the first edition of this book. The prin-ciples of horse breeding have not changed during that time but science, as always, has moved on. Perhaps the most notable advance in relation to horse breeding has been the introduction of ultrasound scanning, to which a new chapter has been devoted. The techniques of artificial insemination (AI) have improved and are being employed more widely in Europe and elsewhere, albeit not in Thoroughbreds where its use remains prohibited. The technology of embryo transfer has also improved in all species and it is being practised increasingly in horses. For these reasons a chapter has been added on AI and embryo transfer. The science of feeding has advanced and a chapter is now devoted to this subject, although there are many excellent reference books available for those wishing to have more detailed informa-tion; a list of relevant further reading has therefore been included.

PART I

THE HORSE AND MAN

1
NATURE'S HORSE

For thousands of years horses have been linked to mankind, as beasts of burden and in war. In more recent times, with the advent and development of the internal combustion engine, the emphasis of use has been on sport and recreation. Nowadays, horses are companion animals, providing people, as riders or spectators, with such pleasurable pursuits as hacking, hunting, showjumping, racing, trotting, three-day eventing and dressage. Every day throughout the world, thousands upon thousands of men and women devote their energies to breeding, rearing, maintaining and working horses of one breed or another. They persevere, in what is often an arduous commitment, for their own satisfaction or gainful employment. The result of their work provides entertainment for millions of onlookers who have only a passing interest in horses and little contact with them, except as spectators round the arena or as viewers of the television screen. The horse has become a national pastime in many countries, and the ancillary industries that serve the many equine activities provide job security and income for many individuals employed in catering, journalism, advertising, gambling networks, farriery, saddlery, transport by air, road and sea and, of course, in the veterinary profession.

All these activities, directly and indirectly connected with the horse, centre on that one species, *Equus caballus,* with its variety of breeds – Shire and Shetland, Thoroughbred and Arab, pony and Appaloosian, etc. The modern horse has a long history of evolution stretching back through aeons of time to small terrier-sized ancestors and, in contemporary times, ranging across to species of wild horses, asses and zebras.

It is important to appreciate the privilege and benefits that nature has bestowed on us by providing such an adaptable companion animal for human use. However, in taking this amiable, enduring and graceful creature from its natural habitat into our civilised way of life, we have to accept not only the benefits of its particular physical and physiological make-up, but also the limits within which these attributes can be used. Motorcars can be of different shapes, size and performance, but we do not expect one of such vehicles to fly; and if it could it would not be a motorcar but some other vehicle. This analogy applies to the horse. Nature has developed a given structure which we cannot change except in terms of size, shape and performance; if we could somehow change the evolutionary scale and produce a horse with a radically altered structure, eg with wings or an extra pair of limbs, it would no longer be a horse. A horse is a horse is a horse, and we have for our purpose only one species, *Equus caballus,* the product of evolution. This species has 64 chromosomes (the threads carrying the genetic material) compared with other horse relatives such as the mountain zebra (32 chromosomes), ass (62) and Przewalski (66). The physical and other characteristics of horses are the expression of their genetic material and can only be altered by selection within certain areas of size and shape, thus providing us with the different breeds. By selective breeding, we can manipulate the characteristics to suit the differing ways we wish to use our horses, eg for riding, racing or pulling heavy loads; but we cannot make fundamental differences in biological functions, such as sex, breathing, digestion or excretion. We may, of course, influence these functions by changing the environment, naturally or artificially. For example, we may breed horses in a hot, dry climate or in a colder and more temperate climate. And we can, to some extent, alter the geographical environment by providing shelter and warmth in cold conditions or by irrigating hot arid regions.

We have to bear in mind, in breeding horses, the influence of the variables of environment, selection for breeding stock and the capacity of the species to adapt within the limits of its inherent characteristics. For example all horses, of whatever breed, have an inherent sexual rhythm that enables them to breed more readily in late spring and summer. This inherent breeding pattern, based on the duration of daylight, was necessary for the survival of the species in evolutionary terms, ensuring that foals were born in favourable climatic conditions. A harmony developed between selection and environment, the one complementing the other. However, we have changed the evolutionary rules and we now select stock for breeding on the basis of specialised characteristics of racing performance, for instance, and we alter the environment by providing spring conditions in winter. We have, therefore, tilted the balance between inherent sexual rhythms and the environment, so that mares breed at any time of the year, irrespective of the hours of daylight or of climatic temperatures. Further, we may actually change the inherent capacity of a given equine population by selecting from those individuals who conform to a desired pattern, say, of breeding during February in the Northern Hemisphere. We cannot, on the other hand, alter the basic nature of the sexual rhythm of the mare, ie the glandular make-up responsible for the behaviour and physiological changes of the oestrous cycle (see chapter 2).

BREEDING A HORSE

The most important consideration in breeding horses is to select individuals for breeding on the basis of helpful characteristics such as fertility and fecundity – the ability to conceive, carry, deliver and rear foals. Next, the breeding stock should be maintained in a suitable environment. Under reasonable conditions of nutrition the mares exhibit sexual activity and accept the stallion in the late spring and summer months. And, eleven months later, foals will appear, if not from all the mares, at least from a substantial number of them. To improve the chances of obtaining a foal from every mare in the group the aspiring breeder should remove from the group those mares that failed to conceive in the first year, and apply the same principle of culling in subsequent years. The same rules of selection should be applied to the male horse and, if subfertile, he should be replaced by another stallion.

By following this route, a breeder will be rewarded with a high percentage of live foals each year. Sufficient land over which to roam must be provided in order to avoid heavy concentrations of parasites, which could cause debilitation and death in many individuals and reduce the breeding efficiency of the group. The wise breeder ensures that pregnant mares have access to good quality spring pasture, especially in the last months of pregnancy, and that these conditions will be available after foaling, when sexual activity and re-mating take place.

Such guidelines enable horse breeders to achieve their objectives. What we cannot guarantee is that the type of horse resulting will be suitable for a particular purpose. The breeder may start with a shape and size of his choice, for example a Shetland, Shire or Thoroughbred, and he will be able to maintain this form indefinitely. But as soon as other selection factors are introduced – a specific colour marking, shape of head, or speed – the natural equation becomes distorted, and sooner or later untoward problems develop. In practice, to concentrate on a limited number of typical characteristics, excluding others, entails selecting from one pool of genes in preference to others. Thus if we select on the basis of the racecourse test for speed and stamina, to the exclusion of factors such as fertility and good mothering qualities, we may inadvertently increase the risks of infertility, poor growth rates and unsoundness. We need not assume that all horses with above-average performance are subfertile or that all highly fertile horses are slow. However, the link is there in a proportion of the population, and if we select entirely for one set of characteristics we will encounter a corresponding increase in the number of individuals with breeding problems. When a breeder has to serve a particular market he must attempt to strike a balance which takes account of fertility as well as of performance.

SOCIAL ORGANISATION AND REPRODUCTION

The contemporary horse breeder should have some knowledge of the ancestral wild horse to understand the problems of breeding. Of course, there are few truly wild species of horse available for study, and many of these are now to be found only in parks and zoos. However, for the purpose of study and understanding we need not travel to Mongolia to look at the Przewalski, or to Africa to observe the wild ass or the zebra; we have only to visit the New Forest or Exmoor to appreciate the social order passed to the modern horse and on which our efforts at breeding are based.

The social and sexual organisation of herd life is important because it is instinctive and inherited. It determines much of the sexual behaviour of present-day horses. Although we do not, and could not for practical reasons, allow this organisation to have full expression, we cannot deal with certain problems on studfarms unless we are at least aware of its significance. Herd life represents the motive behind the behavioural patterns we encounter in our day-to-day duties, whether as vets or as studmen.

There are basically two types of sexual organisation. One, acquired by the horse family, zebra and mountain zebra, is characterised by non-territorial family groups composed of a stallion and a few mares with their foals. These groups (harems) retain their integrity over many years. The young males eventually leave the group to join bachelor groups, from which individuals break away to form new harems. These are composed of young maturing fillies, as they too leave their original group. Some older mares that have become stragglers from their original harem may be incorporated into the new group. Mating occurs between the family stallion and his mares. The presence of the stallion is respected by other stallions, so that there is little or no fighting. The foals remain in the family until they are about two or three years old, leaving more of their own volition than by being driven out by their elders. However, the process of separation of foal from mare and of the yearling or the two-year-old from the group is a delicate balance of positive rejection and voluntary egress from the herd. In this way, all the members of the group enjoy continuous security within the herd; and the herd can travel as a unit over long distances with no fixed territorial limits.

In contrast, a small number of horse species, such as Grevy's zebra, a

A stallion with his mares and their foals on Exmoor. The stallion has come onto the scene, suspicious of the photographer and anxious to protect his mares from outside interference

native of Somaliland and northern Kenya, and donkeys, have a rather different organisation, with no permanent bonds between any two adult individuals. Members of these species may live alone or in association with others, but the composition of the group may change on an almost hourly basis. The only permanent relationship is between the mare and her foal. Further, the stallions are tied to territory which they defend vigorously against other males of their species.

The horses we own and breed are those possessing the first of the two social organisations described above. However, under conditions of modern stud management, we impose on our horses an organisation more typical of the second kind, ie that of Grevy's zebra and the African wild ass. Thus, we place the stallion in a situation which is reminiscent of a male dominance within territorial boundaries, serving a herd of constantly

15

changing composition, rather than the cohesive one-male family group which is characteristic of the non-territorial ancestors of our present-day horses.

The man-imposed social order of the modern studfarm, in most parts of the world, isolates the stallion from his mares, confining him to a loose box and in specially prepared paddocks. It used to be the practice to separate the stallion from any contact with other horses, but nowadays he is allowed rather more freedom and may be exercised by being led or ridden round the studfarm; thus the stallion has some distant contact with the mares in the herd. However, he is never allowed to, nor need he, dispute the territory allotted to him within the organisation of the studfarm.

The mares run in groups which are quite arbitrarily selected according to such criteria as whether or not a mare is barren, pregnant or has a foal at foot. The groups come together for relatively short periods and are certainly never permanent beyond a matter of weeks or months. Members of the group are further isolated and re-arranged by the habit of bringing the mares into loose boxes at night. In the Southern Hemisphere, in Australia, New Zealand and South Africa, mares are rarely confined to loose boxes at any time of the year. Instead they are allowed group contacts which are nearer to those enjoyed by their ancestors and by wild horses, albeit without enjoying the presence of the stallion. It is only in comparatively exceptional circumstances that mares are run with the stallion under free-ranging conditions.

The least artificial management conditions are those of semi-wild herds which roam over moors and through forests. The organisation here is very similar to the natural cohesive one-male family group with the stallion having exclusive mating rights. We cannot compare this almost natural situation with the more intensive management of Thoroughbred stallions that may be run with mares during the breeding season for the purpose of free mating. In these circumstances, the herd consists of members whose behavioural patterns have been 'civilised' by breaking, riding and other means; and the composition of the group is temporary, being entirely devoted to the purpose of the breeding season.

The social organisation of the natural harem and the bonds which develop between members of the group – the stallion, mares, foals, yearlings and young colts and fillies – are based on the needs of mating, genetic inheritance and feeding. The stallion is dominant within the harem and the male/female bonds establish a long-term loyalty that lasts throughout the lifetime of the mares. Sexual signs are interpreted by the stallion, who sees the stance of the mare straddled in oestrus as a visual invitation to investigate her sexual state. However, the odour of a mare in heat is a more certain signal and one which carries some distance. Finally the stallion confirms the sexual state by taste. Thus the harem represents a group that can wander over long distances in search of food but at the same time retain a cohesive relationship, with mating occurring when the mares are sexually ripe for the occasion. This system has developed through the evolutionary forces of selection, providing the best characters for optimal chances of survival. Under natural conditions, selection-favoured stock bred only in the spring and summer months, which determined that the birth of their foals a year later would be in the season of adequate nourishment and clement climatic conditions. The ability of the group to migrate, ie those that were non-territorial, ensured that they were not restricted to an area that might become arid and devoid of food and water.

The genetic pool was conserved by the harem system and not dissi-pated throughout a wide range of individuals. The effect of this restriction meant that successful groups, ie those with genes responsible for charac-ters of optimal survival in a hostile environment, were preserved and maintained their position in the herd, relative to less successful harems. Success bred success, a feature of the mechanism of evolution to which all living creatures are subject. The mixing of one pool of genes with another and the creation of new gene pools was, of course, catered for by the young stallions developing new harems of their own, the members of which were stragglers separated from established harems or young fillies leaving their harem on the 'coming of age'. The departure of the young also limited the amount of inbreeding that might otherwise have occurred within the harem as father mated daughters. It also avoided the risks of enlarging the harem to numbers which could not be adequately satisfied by one stallion.

We can discern fragments of this social organisation and reproductive behaviour – evolved over many millions of years – in our horses on the studfarms of today. We have, by management, interfered with the process of natural evolution, by altering the environment and selecting for our particular purpose, depending on the breed. We can marvel at the way the species has co-operated in the process of change. The modern horse has adapted to being confined to territories, the members of a harem separated and isolated in a haphazard fashion, and the breeding season changed so that mares are now mated and foal in winter, as well as in spring, summer and autumn. We have paid the price for these alterations in having to supply food and protection to correlate with this change in mating and foaling habits. Thus we are condemned to making hay and storing fodder for feeding high levels of energy and protein; to building expensive loose boxes in which to house the breeding stock and shelter them from adverse climates at times when spring and summer conditions are required. The measures needed to simulate spring in winter are expensive indeed; much labour and effort are required to produce the necessary feedstuffs and maintain the environment of the mares confined to loose boxes and limited areas of pasture which must be kept relatively free of parasites. In the natural state, none of these measures is required.

The biological background to the social and sexual behaviour of horses is the glands and sexual organs of the body. Our success in manipulating these glands and organs and moulding their function to our requirements depends, to a large extent, on our ability to understand the biological processes involved. The purpose of this book is to explain to the reader the basic biological principles of structure and function that control the reproductive capacity of mares and stallions in health and disease. It is only by understanding the complex systems and their interactions within the body that we can interpret the outward manifestations of equine reproductive behaviour and control them to our advantage in economic terms. The breeding of a horse is simple if management is left to nature; but when *we* are the managers, it is *we* who have to comprehend what we are doing if we are to succeed; many of the problems and diseases that we encounter are the direct result of too little understanding. It is probably true that the further we move away from nature's management, the more difficult our task becomes in breeding horses, the more problems we encounter and, therefore, the more important it is that we understand the biological mechanisms of the 'machine' we are driving.

2

VETERINARY SCIENCE
AND MANAGEMENT

Vets have become increasingly drawn into the horse breeding industry during the last quarter of a century. The reasons are not hard to find; vets are trained in the science of reproductive function, and scientific knowledge expands at an ever accelerating rate. Horse breeders, on their side, face rising costs and capital expenditure necessary for the purchase of mares and stallions. The value of land for grazing and growing fodder, the salaries of studfarm attendants and farm workers, the price of transport and taxes have all escalated. In this climate, the incentive to increase productivity and thereby to decrease unit costs is considerable; but, at the same time, there is pressure to meet market demands for such artificial endpoints as conformation, size and early birthdate. All aids seem therefore to be justified, and it is not surprising that breeders have turned to veterinary science and veterinarians for help with their problems. Within the pages of this book we must examine the extent to which their trust in science has been fulfilled, or even justified. On the one hand we have more horses, decreased acreages of pasture available for breeding horses, limited stabling and expert manpower, served by reduced farm output of cereals and hay, resulting in overcrowding and inferior feeding; on the other hand, management has access to more sophisticated techniques, therapies and preventive measures developed in laboratories, veterinary practices and institutes.

It is difficult to make an objective assessment of these considerations in terms of success and failure. In most Thoroughbred stud books, the records show an increased productivity in terms of total numbers of live foals born each year. In the first edition of this book figures between 1957 and 1979 from the Thoroughbred *General Stud Book* of the UK were used to show that the total number of mares increased to a peak of 17,500 in 1975 from a low point of 7,500 in 1957. During this period, the percentages of mares in foal and of live foals remained static. More recently there appears to have been an improvement (fig 1) in the percentage of mares in foal. The reasons for this are not clear because there have been no objective studies on the subject and, in any case, statistics reported in general stud books rely on the returns provided by the breeders which, in themselves, may change according to fashion and circumstances. For

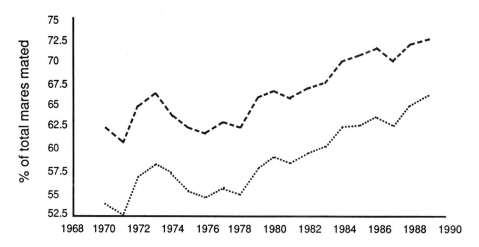

Fig 1 Conception (- - -) and live foal (······) rates as a percentage of total mares mated between 1970 and 1989

example, a mare tested in foal at, say, sixty days from last service but found to have lost that pregnancy after a further three months may be reported as having slipped or as being barren. Nevertheless, an increase in productivity (percentage of live foals) does appear to have taken place. This may have been the result of owners culling brood mares with poor breeding performance due to economic constraints; to improved managerial and veterinary skills; or to a combination of these factors. Further, results obtained from very large samples do not reflect what can be achieved in smaller groups subjected to optimum conditions of management. There are many variables that must be taken into account in assessing the contribution of science to the horse breeding industry. It can be argued that more attention should be paid to establishing ongoing statistics, based on carefully collected information, that could be used to compare progress, if not from year to year, at least from one decade to another. It would seem to be an admirable function for breed societies to introduce and control these studies.

We have already discussed the effects of selection on fertility, and it is clear that if breeders were to select mares with a good breeding record, and were to discard those with inferior records, the situation in any breed of horse would improve, with or without veterinary assistance. It is in those breeds, such as the Thoroughbred, where breeders refuse to adopt the simple policy of culling inferior breeding stock from their herds, that we need veterinary assistance to overcome the deficiencies faced by breeders in consequence of their own actions. In the real world of breeding horses for high-powered performance – trotting, racing, jumping and three-day eventing, etc – management often requires science as a shield against the problems that their policies entail, including the effects of overcrowding, inadequate quality of fodder, and selective breeding within arbitrarily selected calendar months for artificial ends and early maturity. All of these effects culminate in disease and conditions of disordered growth and injury. In the words of the popular song, you can't have one without the other; intensive management and husbandry are inevitably coupled with disease and disorders; in consequence, veterinary science and vets are much involved in giving advice and assistance.

PROGRESS IN VETERINARY SCIENCE

Hormonal Function

Veterinary science has much of benefit to offer horse breeders. The 1950s were the watershed from which significant progress developed with increasing application to equine veterinary practice. Prior to 1950 the most important contributions of science to studfarm management were a knowledge of the oestrous cycle of the mare, gained by observing her behaviour (page 81), and relating this to information obtained through rectal palpation of the ovaries and the examination of the cervix by means of a speculum. These techniques laid the basis for the routine gynaecological examinations which have become such a feature of post-war studfarm management throughout the world. It was these examinations which placed the veterinary profession firmly in the centre of stud management, largely taking over the responsibility of the studgroom at the trying board, because of the vet's ability to predict the optimal time for mating, and to diagnose oestrus and pregnancy. These matters are discussed more fully on page 71.

The hormonal events of the oestrous cycle, and of pregnancy, were little understood before 1950, with two notable exceptions. These exceptions were (1) the ability to cause ovulation of a ripe ovarian follicle by administering a luteinising hormone (LH) such as HCG (human chorionic gonadotrophin), which is obtained from the urine of pregnant women; and (2) the development of endometrial cups (see page 143), saucerlike structures in the pregnant mare's uterus, as the source of a powerful pregnancy hormone (PMSG). New techniques for measuring hormones and for investigating biological changes in the body have provided an increasing understanding of the mare's oestrous cycle. This new knowledge has been accompanied by extraordinary progress in synthesising drugs which

The vaginal examination with a speculum is a means of examining the cervix for signs of oestrus and for collecting material for laboratory examination

mimic the action of natural hormones, often in a more powerful fashion. The combination of knowledge and effective therapy has enabled us to manipulate the oestrous cycle with ever-increasing success. However, these advances have not as yet enabled us to make significant progress in solving the problems of infertility.

In the stallion, the quality of semen relative to the horse's fertility has received some attention, but unfortunately our knowledge and understanding of horse semen has not kept pace with that of bull semen. This discrepancy is ascribed to the banning of the use of artificial insemination (AI) in many horse breeds by the stud book authorities. This has prevented a detailed examination of stallion semen, and information has only become available through certain breeds, such as Standardbreds, where AI has been permitted. In practice, one of the problems encountered in the examination of stallion semen is the very high number of abnormal forms which are found, even in horses of good fertility. For this reason it has been particularly difficult to forecast the fertility of a young horse on the basis of one semen examination; and there have been a substantial number of cases where such forecasts have been proved, subsequently, to be inaccurate.

The diagnosis of the cause of infertility often remains obscure, even today, despite modern technological advances. Nor has treatment improved even in those cases where the diagnosis of the problem is certain. It may well be that the solution to stallion infertility lies in avoiding the use of subfertile stallions which have a genetic susceptibility to infertility.

Infection

Microbial infections were well recognised in the 1930s as the cause of venereal disease associated with infective conditions of the uterus and genital tract. American scientists in Kentucky described the troublesome nature of *Klebsiella,* streptococcal and other infections, and made some progress in treatment by developing the surgical operation on the vulva to eliminate pneumovagina (see page 249). The condition of viral abortion was also first described in the USA and subsequently identified as occurring in most, if not all, countries in which horses are bred. Epidemic storms of abortion sometimes occurred, and this disease, much dreaded by breeders, is still present with us today. Although we understand more about it than in pre-war years, there remains much to learn about the way it spreads and the means of its prevention. The disease is now known as rhinopneumonitis and the causal viral agent as Equid herpesvirus I (see page 270).

Personality Cults

The veterinarian before the 1950s was restricted by a lack of knowledge and understanding of many of the processes and conditions encountered in practice. Further, modern drugs such as antibiotics and synthetic hormone preparations were not yet available. In those days the clinician relied more on direct observation at the 'patient's bedside' than on laboratory, radiographic and other ancillary aids. Professional reputation depended on judgement based on experience; a vet who was steeped in horse tradition and terminology had a considerable advantage over less-familiarised colleagues; and the riding and hunting vet was likely to be held in greater esteem by the laity. The veterinary training of twenty-five years ago equipped a graduate with a discipline of thought and action, which is the hallmark of a scientific attitude. However, science itself was but a pale reflection of present-day possibilities, and those who practised the art of equine clinical medicine and surgery worked within strictly limited horizons of diagnosis and therapy. The opinion of the practitioner carried weight and authority provided he (it was rarely, if ever, she) had gained sufficient reputation as a horse vet to have become established and accepted by the veterinary surgeon's most critical clientele, the horse owners.

The consequences of a high degree of personal expertise and a relatively deficient scientific background were to exaggerate the personality cult. This cult affects all professions in one way or another, but it operates to its greatest extent where opinion and experience are at a premium and expertise, in the sense of factual knowledge, is limited. Young aspiring members of a profession are often the casualties in these circumstances because they cannot compete against the older, experienced graduate on an objective basis. An establishment 'magic circle' tends to operate in all walks of life where a closed community exists, and the veterinary profession is no exception. However, as the variety and usefulness of scientific methods have increased over the years, so young graduates have been presented with new avenues of opportunity; and the demand for veterinary services has correspondingly broadened.

Diagnosis

Modern developments in instruments and techniques are now available to every practitioner. For example, radiography was once the privilege only of veterinary schools and institutes, but nowadays machines of considerable power and efficacy can be readily transported into a loose box in order to X-ray a damaged part. Portable X-ray equipment, capable of photographing the limbs, is now within the reach of all practices. Many practices now have machines capable of producing radiographs from more substantial parts of the body, such as the neck and back. Other forms of diagnostic equipment now in common use include flexible fiberoptic endoscopes, which provide a direct view of the upper airways of the head and throat; and similar instruments may be employed for viewing the interior of the abdomen. Laboratory techniques have become more refined and the interpretation of results more precise, as factual data have become available through increasing research and use of these methods in practice. Diagnosis has consequently developed into more of a science than the art it was some thirty-five years ago. During the past ten years ultrasound scanning, videoendoscopes and laboratory analysis of hormone concentrations based on radioimmune assay (RIA) have widened enormously the scope of diagnosis in all branches of veterinary medicine.

Therapy

Corresponding improvements have occurred in medical and surgical therapy. The range and effectiveness of drugs have increased dramatically, bringing benefits undreamt of fifty, even twenty-five, years ago. Antibiotics have completely altered our management and treatment of infectious diseases and of simple conditions such as abscesses and dermatitis, some of which are serious and some annoying, but all bringing economic consequences. Anti-inflammatory drugs, eg phenylbutazone (bute) and cortisone, are used to great effect in the control of sprains,

An endoscopy being performed by Tim Greet of Beaufort Cottage Stables, Newmarket

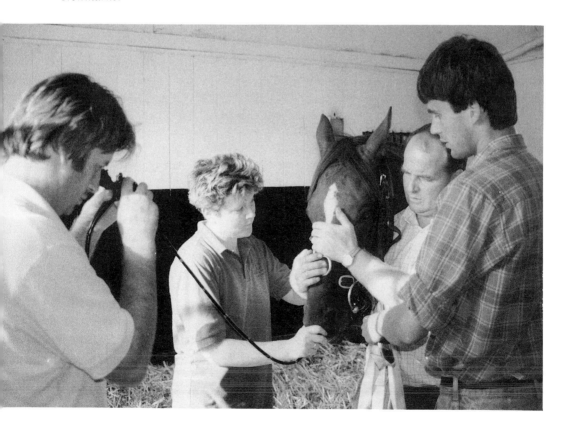

injuries, 'filled legs' and lameness. Vaccines against influenza and herpesvirus (rhinopneumonitis) have been added to our armoury in the fight against the spread of epidemic infections.

General anaesthesia was once considered a risky enterprise, especially for surgical procedures lasting any length of time. Indeed, major abdominal surgery (caesareans, twisted gut) or orthopaedic surgery (bone and joint) were seldom attempted. However, when gaseous anaesthesia, developed for humans and small animals, was successfully adapted for equine use, a new era of equine surgery began.

Standard of Service

These innovations in veterinary science enhanced the services that could be offered to the client. Gradually, the horse owner has come to expect an increasingly higher standard of service incorporating modern diagnostic techniques and therapy, to be used on horses of ever-rising money values. More and more graduates are required to occupy these new areas of endeavour as the frontiers of veterinary science have extended. The oligarchy of the old order has been replaced by a much broader spectrum of young graduates, whose reputation and expertise have become established throughout the country; and not in a few selected areas only, such as Newmarket, Lambourn, Leicestershire and Yorkshire where the concentration of horses is greatest. Equine practice has now become a speciality within the veterinary profession, and most practices have a specialist familiar with the horse.

Other influences on professional progress include the wider dissemination of knowledge and experience through such organisations as the British Equine Veterinary Association. BEVA was established in 1962 under the chairmanship of Lieutenant-Colonel John Hickman of the Cambridge Veterinary School. Membership has risen to include nearly all practising equine clinicians, as well as representatives of the veterinary schools, equine research stations and other scientific institutes. Congresses and meetings are attended by delegates from all parts of the world, reflecting the broadly based contact that vets now have within their own and foreign countries. BEVA publishes its own *Equine Veterinary Journal* and, more recently, *Equine Veterinary Education*. The association has undoubtedly made a significant contribution to the trend providing the profession with a large measure of equality of status, and not dominated by a select few residing in key situations. The equine side of the profession has thus become more inclusive than exclusive. The consequence of this more democratic approach has been to enhance enormously the capabilities of individuals, professional prestige and the quality of services that can be offered to the community.

Nor is BEVA the only organisation that has aided the liberal process of scientific advance. Most countries have followed the lead of BEVA and its American counterpart, the American Association of Equine Practitioners (AAEP), by forming their own equine associations. Further, in the field of equine reproduction, a new professional body was set up in Cambridge in 1974 to arrange meetings every four years. This is now known as the International Equine Reproduction Symposia Committee. The object of each meeting (symposium) is to bring together the leading scientists of the world, investigating problems of equine breeding in the widest sense, from the stallion to the mare and from the foetus to the newborn foal. Subsequently, symposia have been held every four years: at Davis, California in 1978; Sydney, Australia in 1982; Calgary, Canada in 1986 and Deauville, France in 1990. The proceedings are published as supplements to the *Journals of Reproduction and Fertility*.

The net effect of professional progress in clinical medicine, surgery and research over the last two decades is that veterinary science has considerably more to offer by way of diagnostic and therapeutic services than was the case in previous years; and the trend continues at an ever-increasing pace. It is now possible to measure almost any hormone in the mare's blood stream and to follow changing patterns through the breeding cycle. The term hormonal imbalance, so often used in the past without any scientific basis, can now be proved in objective terms by assessing the levels of hormones responsible for the oestrous cycle and the maintenance of pregnancy. It is one of the main purposes of the present book to explain to readers the meaning and possible practical applications of these new discoveries.

SIDE EFFECTS OF SCIENCE

However, as with all scientific progress, there is another side to the coin. We cannot tamper with nature without bearing some of the penalties. There are millions of people alive today who have been helped by antibiotics; but there are some whose health has been temporarily, or even permanently, damaged by their use. It is important for veterinarians and horse owners to recognise the harmful, as well as the beneficial, consequences of applying science to horse breeding. It is not only drugs which may have untoward repercussions, but the influence of science on managerial methods. There are, for example, mares and stallions whose breeding life depends on the aid of veterinary science; these individuals

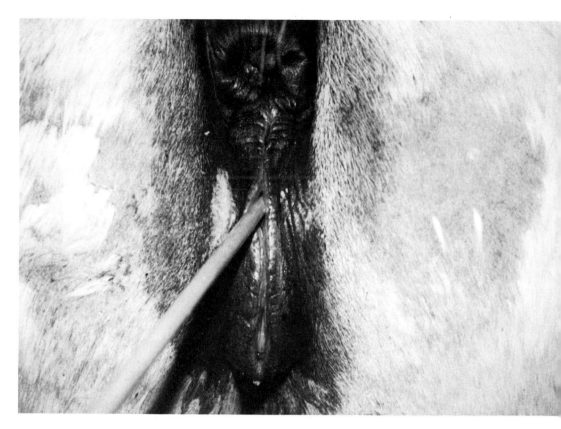

Good conformation of the structures below the base of the tail is an important attribute in sustaining fertility of mares. A metal rod is seen resting on the floor of the vagina and the greater length of the vulva is, normally, below this level

would in earlier times have been eliminated (culled) from the herd. Some of the conditions necessitating veterinary treatment have an inherited basis. In nature, selection operates to eliminate these defects; the individuals with a tendency to infertility may breed progeny with increasing difficulties and, eventually, some that are sterile; thus the defective characteristics are automatically removed from the herd. But if we artificially protect subfertile individuals and promote their breeding capacity we ensure the propagation of defective genetic characters, and an increase in the number of individuals within the herd requiring veterinary attention. Not only are there more individuals requiring attention, but each succeeding generation may become progressively more dependent on scientific methods.

Let us consider how this unpleasant reality may apply in practice. In the condition of pneumovagina, described on page 249, air abnormally enters the vagina owing to a weakness in conformation of the anatomical structures below the base of the tail. This weakness is an inherited character. Thus, if we select individuals for breeding with this type of conformation we produce individuals with a weakness as great as, if not worse than, that of their parents and grandparents. The consequence of air entering the genital tract is to produce infection and subfertility, ie a diffi-culty of getting 'in foal'. If the condition is left untreated the damage to the uterine lining is so great that the individual becomes completely infertile, ie sterile. Dr Caslick, a veterinarian in the USA, solved this particular problem by developing a surgical operation to correct the anatomical defect. This operation, first performed in the 1930s, has effec-tively restored the breeding life of many thousands of mares and prevented them from becoming subfertile and sterile. Yet the success of the procedure has, in itself, ensured that more individuals in the popula-tion suffer from the anatomical defect and that the degree of abnormality has tended to increase within the population. Nowadays we have to 'stitch' more individuals to a greater length of the vulval aperture than was the case twenty years ago. In a survey on two Thoroughbred studfarms in Newmarket, thirty-three per cent of mares were found to have been 'Caslicked'.

Antibiotic therapy is a further example of untoward happenings. Antibiotics can often effectively deal with infection, and it is not surprising that these substances – penicillin, neomycin, gentamicin, ampi-cillin, to name but a few – have been used extensively in veterinary practice for the treatment of infective infertility in mares and stallions. Many mares have been cured and stallions freed of harmful microbes, but in the process two undesirable side effects have been recorded: (1) microbes (bacteria and fungus), resistant to the drug being used, may flourish at the expense of their harmless competitors. In this way, the natural inhabitants of the penis and the mare's genital tract become permanently changed and a new population of microbes is established. For example, a microbe known as *Pseudomonas* may colonise the penis of a stallion that has been treated with antibiotics or antiseptics in an effort to eliminate the highly contagious CEM (contagious equine metritis) microbe; (2) antimicrobial treatment may produce strains of microbes which are resistant to the drug of choice, and it is known that microbes may transfer the resistance they develop to other microbes. Thus the treatment of our choice may fail to eliminate an infection.

SCIENCE AND METHODS OF MANAGEMENT

In recent years the management of horse breeding has been radically changed by the intrusion of veterinary science. A major impact occurred in the late 1940s and early '50s, when the techniques of ovarian palpation and vaginal inspection were introduced into veterinary practices serving the major Thoroughbred breeding areas of the UK, Ireland and the USA. The objective of the routine veterinary examination, as it has become known, was identical to that traditionally undertaken by the studgroom or stud manager, namely to determine which individuals were in heat on any given day and when these individuals should be mated by the stallion. Studmen employed, and still do employ, the teasing board for the purpose of diagnosing oestrus, but the veterinarian has become increasingly relied upon to confirm the mare's sexual signs displayed at the teasing board. As

Mare in oestrus showing antagonistic signs at the teasing board instead of usual signs of oestrus

at all frontiers between human management and animal biology, the demand for artificial services has risen to meet their availability. In the context of horse breeding, more mares nowadays require veterinary-aided diagnosis than hitherto. This is partly due to the attitude of stud personnel, who find it more convenient to have a vet make a diagnosis, with a high degree of certainty, than to rely on their own observations, which are necessarily limited by the subjective deficiencies of the method. Further, the studman relies on the mare to show genuine sexual signs when stimulated, ie to inform by her behaviour that she is in oestrus. But for various reasons increasing numbers of mares fail to co-operate and thus do not display the expected signs at the teasing board.

Twins

Routine gynaecological examination was originally justified by the claim of early workers that twins could be avoided by palpating the ovaries and determining whether or not two follicles were present. In the event of two follicles being detected, it was argued that mating could be arranged after one, and before the other, follicle had ovulated, thereby avoiding the conception of twins. These early hopes were not realised and Thorough-bred twins were as common as before rectal examinations were first practised. Roughly two per cent of all Thoroughbred pregnancies (higher in some other breeds) started as twin conceptions and consequently ended in economic failure due to abortion or the birth of an undersized foal. In 1980, the average cost of these pregnancies was estimated at £2,000 per annum and the economic sacrifice suffered by breeders in the UK was estimated to be at at least £400,000 per annum. In actual terms, including nomination fees, capital depreciation and damage suffered by mares after aborting twins, the figure could have been multiplied many times. In the first edition of this book, the problem was stated as one deserving intensive veterinary research. It was pointed out that its solution would be a significant contribution to horse breeding, as well as redeeming the reputation of our profession by substantiating claims made erroneously more than two decades ago. This ideal has now become a reality through the development of ultrasound techniques (chapter 15). When this aid is employed in routine gynaecological examinations, the incidence of twins is reduced almost to zero.

Pregnancy Diagnosis

The diagnosis of early pregnancy is the most substantial benefit that routine veterinary examinations offer management during the breeding season. In the past, studgrooms had to rely on signs at the teasing board over the period of the next expected oestrus following mating, ie two to four weeks, before assuming that the individual was pregnant. They relied on the further evidence that no signs of heat were shown over the period of the second expected oestrus, five to seven weeks after service. Subsequently the mares would be observed for oestrus until the end of the breeding season. This programme was fairly reliable but it was recognised that a number of mares come into oestrus outside the expected period (see page 87). These individuals are said to 'cycle' irregularly. Other individuals undergo long periods (two, three or four months) during which they show no signs of sexual activity. A pregnancy diagnosis in these circumstances is invaluable. The advent of ultrasound scanning (see page 350) has enabled us to provide an accurate diagnosis from about Day 4 onward and the studgroom, therefore, can be informed whether or not a mare is pregnant.

Parasite Control and Preventive Medicine

The influence of veterinary science on management has been even more dramatic in the measures and advice it has been able to offer in the fields of parasite control and preventive medicine. There are a remarkable number of newly available drugs which are effective against the main equine parasites. In particular Ivermectin, a drug capable of killing parasites in the blood and tissues as well as those in the gut, has provided a far more effective means of parasite control and enables us to treat, and not just prevent, infestation. Vaccines and other prophylactic measures against infection have improved, but there is still a long way to go before the risk of debilitating bacterial and viral infections is eliminated.

RESEARCH

The veterinary profession organises its own research resources in such a way that they are harnessed to investigating the really significant problems facing breeders. However, too many projects have been undertaken on an *ad hoc* basis with little, if any, hope of producing practical returns for horse breeders. The steps between the setting up and undertaking of a research project leading to some conclusions that can be applied in practice are numerous, time-consuming, and often very expensive. There are limited resources of money and manpower available for this research, so it is particularly important to organise our efforts in a co-ordinated and logical manner. There are a number of grant-aiding bodies, such as the Horserace Betting Levy Board, the Wellcome Trust Home of Rest for Horses and, most recently, the equine Virology Research Foundation (EVRF), which have over the last two decades contributed substantial sums towards equine veterinary research. Individuals, associations and breed societies have also provided significant support in many instances, but the results have not always met the expectations of the donors. There has been too much research undertaken over a wide field with too little account of its likely endpoint in practical benefits to the horse, the horse owner, or the country's equine industries. But, even more regrettably, endeavours have been too dissipated to have had any hope of solving some of the major problems facing vets and horse managers. The majority of projects are developed as an end in themselves, to satisfy scientific curiosity, to improve the career prospects of the investigator by acting as a means to a higher degree, or to solve a relatively unimportant problem, rarely encountered in the field. All of these endeavours may be justified in a climate of unlimited research funds and the under-employment of graduates. Unfortunately this climate of superfluity does not exist today – in fact, quite the reverse. It would seem more prudent to give some priority to areas of particular importance where substantial benefits might accrue from solving specific problems, eg improving methods of diagnosing the causes of infertility in stallions, developing a vaccine against the redworm parasite *Strongylus vulgaris,* and diminishing foetal and neonatal wastage through a better understanding of its causes.The EVRF has devoted its considerable annual support towards research into Equid herpesvirus infection, the causes of respiratory problems in young horses, and of abortion in mares.

We need, in practice, to know more about the significance of venereal microbes. A great deal more swabbing of the mare's genital tract and the stallion's external genitalia is undertaken nowadays than previously, following the outbreaks of *Klebsiella* in the 1960s and CEM in the late '70s. There has been an inevitable rise in the findings of venereal and potentially venereal microbes, but the significance of many of these organisms is still unassessed. Laboratory and field experiments are necessary before we can solve the problems of interpretation which confront clinicians, stud managers and owners.

There are other problems worthy of consideration, but it would be more profitable for representatives of the veterinary profession and the horse industry to come to some conclusions as to priority and then to harness efforts in a co-ordinated and substantial manner towards stated objectives – instead of relying, as at present, on a haphazard approach.

In 1978 the British Equine Veterinary Association Trust developed a liaison committee with these aims in mind. The committee consists of representatives from the following: Animal Health Trust, British Equine Veterinary Association, British Equine Veterinary Association Trust, British Horse Society, British Veterinary Association, Donkey Sanctuary, Equine Virology Research Foundation, Home of Rest for Horses, Horserace Betting Levy Board, International League for the Protection of Horses, Irish Equine Centre, Jockey Club, Ministry of Agriculture, Fisheries and Food, Moredun Research Institute, National Trainers' Federation, National Light Horse Breeding Society, Racehorse Owners' Association, Royal College of Veterinary Surgeons, Thoroughbred Breeders' Association, University Veterinary Colleges, Wellcome Trust.

AIMS OF THE VETERINARIAN

Veterinary responsibility is not confined to treating the sick and curing disease; vets have a function in preventive medicine and in providing advice on the management and husbandry of horses with a view to improving the health, productivity and soundness of breeding and working stock. But the vet also has a duty to teach younger graduates and to pass knowledge and experience to undergraduates and colleagues. The means whereby the vet communicates to others are lectures, discussion groups, demonstrations and articles written to explain the general and the particular, ie broadly based subjects and the results of original research.

There can be little doubt that the status and progress of the profession depend on the communication of ideas and research results, and subsequent discussion and criticism by the widest possible audience. Truth that is confined, hidden or not communicated is truth wasted and, more important, untested. Professional progress depends on the testing of all assertions, methods and techniques to determine their truth, accuracy and efficacy.

But there is a further professional duty, one to which this present work is dedicated, namely communicating to non-professional people, in this case those who own and care for horses. It is essential that these people understand the aims, limits and endeavours of veterinary science in its broadest sense – as practised by clinicians, as investigated by research workers, and as supported by auxiliary functionaries like drug manufacturers, food suppliers, farriers, and harness makers.

Why, the reader may ask, is it important for lay people to understand biological processes and conditions of disease? The answer lies in the fact that vets and the laity are partners in maintaining the health of horses and in preventing, diagnosing and curing their ailments. The vet depends on the studman to observe and report signs of ill health. It is these signs and the ability of the studman to recognise them which in the first instance establish that an individual is ill when there is still time to start treatment which will alter the course of the condition. The accuracy of the studman's report is essential, in many instances, to the diagnosis and to the conduct of a case. In treatment, the lay attendant's help and assistance is often crucial to the successful outcome of the case. We shall see (page 297) how important is the human care element in the nursing of ill newborn foals.

It is not only in cases of disease that vets have to rely heavily on attendants, but also in areas of management, such as in the teasing of mares and in their restraint during veterinary examinations. Lay people who work with horses are involved at all levels of veterinary expertise, especially in the diagnosis, prevention and treatment of disease. Their contribution is enormously enhanced if they understand the basic principles upon which veterinary actions are taken. Anyone who is present at foaling needs to know how to act; people cannot act properly if they do not understand the process they are observing. For example, many years ago it was customary for studgrooms to seize the umbilical cord as soon as the foal had been delivered, to clamp and cut it. This procedure has been shown to be harmful because it deprives the foal of a large quantity of blood circulating in the placenta (see page 134), and by placing a ligature on the stump left at the navel, tissue is 'strangled', cut off from its blood supply and left to decay, thereby inviting microbes to breed. Studmen are now taught that this is an unnecessary and potentially harmful procedure, although there are occasionally cases where the cord may have to be broken artificially or a ligature applied. The studman must understand what is happening at the time, if he or she is to take appropriate action in the particular circumstances of each foaling. It is not sufficient for vets to dictate what should or should not be done; it is for those present to appreciate the biological happenings; and to base their decisions at the time on their knowledge and understanding of the events.

The use of scientific and technical jargon often prevents people from communicating with one another, and sometimes leads to misunderstandings. Vets have to be at pains to explain natural function to lay people in terms that they can understand. This is not always a simple matter, because by its very nature the subject is inevitably complicated – incomprehensible at times even to the specialist. Lay people, however, do come to appreciate the advantages of jargon and may incorporate some terms into their own vocabulary. For example, *Klebsiella* is the name of a microbe causing infectious infertility of mares and stallions. IL was recognised by scientists in the 1930s as a cause of infertility, but hardly known in veterinary and lay circles until, in 1962, an epidemic occurred in the Newmarket area. Since that time, there has been much debate and comment over the significance and dangers of this particular organism, including a Thoroughbred Breeders' Association committee, named after its chairman, Lord Porchester, which was appointed to investigate the outbreak in the early 1960s. *Klebsiella* has thus gradually become a household word capable of being pronounced by most breeders. However, experience has shown that there are numerous different strains of *Klebsiella* with varying degrees of infectivity and harmful consequence. The discovery of *Klebsiella* in the mare's or stallion's genitalia does not necessarily imply that the individual is suffering from a venereal disease. The term *Klebsiella* needs interpretation, according to its particular nature and the circumstances in which it is found. We cannot simplify the situation merely by using terminology, and we are faced, as in all biological matters, with the conflict between the need for simplicity and the risks of oversimplification. My aim in this book is to steer a course between these two extremes, to keep jargon to a minimum and, where jargon is unavoidable, to explain technical terms for those not familiar with their meaning. The intention is not to intrude into the relationship of readers with their professional advisers, nor to usurp the prerogative of the attending veterinarian to diagnose and treat reproductive disorders, but rather to promote understanding with a view to improving the ease with which each individual can communicate with those on the other side of the professional fence. In no place will the reader find categorical statements concerning diagnosis and treatment. It is in these areas that professional men and women need the greatest freedom to decide for themselves in the light of their own experience and the particular circumstances of the case.

The world of horses is full of romantic feelings. To a varying extent we all like to identify with these magnificent four-legged beasts. Our attitudes are often coloured by a natural wish to introduce human emotions into the 'equine mind'. Thus, a nice warm bran mash is the equivalent of a hot meal after a day in the hunting field; and the fuggy, straw-bedded loose box represents the favourite armchair beside the blazing hearth. The mythology of horse breeding occupies the middle ground between a strict scientific interpretation and an entirely natural environment. We cannot nowadays rely on the methods of natural breeding, for obvious reasons; small harems running with a stallion are uneconomic, nor would they lead to the highly specialised characteristics we require in our working horses. To achieve our ends managerial control must dominate the natural processes. For example, economic reasons dictate that the number of mares in the harem served by a stallion should be increased from the natural 5 to 15 up to 45 or more. A further modern-day requirement is that the actual membership of the harem changes annually to meet the dictates of the breeder's programme, whereby individual mares and stallions are selected in an attempt to breed progeny of a type to meet market, working and other requirements. The consequence of these departures from natural conditions has been for breeders to turn to veterinary science to overcome the problems that artificial systems inevitably introduce. Veterinarians are now involved in every sphere of horse breeding, from assisting in management decisions regarding mating to the diagnosis and treatment of infertility; and from the control of infectious diseases to advice on nutrition, pasture management and parasite control. The responsibilities of the veterinarian on the studfarm are thus numerous and variable, as the reader will appreciate from a perusal of Parts II and III of this book. The question that must be asked at regular intervals is whether or not our profession is succeeding in both meeting breeders' requirements and increasing the productivity of the industry. In the war there was a slogan 'Is your journey really necessary?' It is appropriate that vets, who have become so closely associated with stud management and horse breeding, should be subjected to a similar scrutiny implied in the question, 'Is your examination/treatment/advice really helpful or necessary?' It is only by a critical appraisal of our actions that we can hope to modify our approach to meet the needs of breeders. But it is equally important for breeders to analyse the logic of their own aspirations and the demands they make on vets and veterinary science. The attitudes of breeders and veterinarians each affect the actions of the other.

Breeders determine the dates of the breeding season and seek the help of vets to enable them to breed foals out of the natural season, ie January and February in the Northern Hemisphere and August and September in the Southern. Veterinary science has expanded to meet this 'unnatural' demand, through research on such aids as artificial lighting and treatment with hormones (prostaglandin, progesterone, releasing factors; see pages 101–2). The veterinary measures have been quite successful but we cannot yet effectively deal with a state of anoestrus. In this state the mare is sexually inactive and her ovaries cannot be stimulated to activity by artificial means. Thus veterinary science is at present unable to meet the breeders' demands for matings early in the breeding season in every case. We must be watchful that by achieving one objective in a given area we do not make problems in another; and that by artificially stimulating sexual activity in the winter we do not make it more difficult to get the treated individuals in foal later in the breeding season, should they not conceive earlier. The old adage that if you sup at the devil's table you must eat the devil's brew is somewhat apposite applied to scientific methods and animal breeding.

LEARNING BY COMPARISONS

It is said that comparisons are odious because inevitably they place one party at a disadvantage to the other. But comparing results obtained in differing situations and circumstances can be informative and helpful in all walks of life. This is the basis of the scientific method and it applies equally to veterinary and managerial expertise in horse breeding. Vets may learn by watching or assisting their colleagues performing medical or surgical techniques that differ in style or detail from those they themselves employ.

Horses are bred and reared under extreme climatic conditions with equal success, judged by percentage live foal or conception rates. There is little to choose between the optimal results obtained on studfarms in, say, Newmarket, UK, and Hunter Valley, Australia. Both these regions possess heavy concentrations of Thoroughbred breeding activity with an abundance of managerial and veterinary expertise, and they are also areas in which maximal capital resources are available and the quality of blood-stock and standard of facilities are at their highest. We might therefore expect to experience the best results, in terms of breeding efficiency, from establishments within these two centres of horse breeding.

Climate

In Australia the summer months are dry and hot, to the point that pasture is brown and burnt, whereas in Europe climatic conditions are consistent with green pasture throughout the summer. The climate in both continents enables horses to be unhoused during spring and summer, but in Australia shelter from the midday sun, and from flies, is more necessary than in Europe. Green feed virtually disappears from Australian pastures in the summer unless these are irrigated, which nowadays is often the case. In winter, the contrast of climatic conditions between the two continents is rather greater; Australian weather conditions are generally much less harsh than those in European countries. In Europe weather conditions may at times be extremely harsh, and housing horses in the winter becomes unavoidable. In Australia mares, foals and weanlings may

In the Northern Hemisphere conditions, where horses are housed, attendants are needed to lead between the stables and the paddocks; however, this makes the young stock used to handling

usually be kept outside day and night. The same contrasts of climate and pasture conditions are experienced between Canada and South Africa or New Zealand; and in the USA between northern and southern states. Variations in the range of temperature, rainfall and associated environmental conditions may occur within one country according to elevation above sea level; indeed, within one country, it is possible to find extremes generally experienced between countries or continents.

Effect of Climate on Managerial Methods

We are concerned here not so much with the differences in climate as such, but with their effect on managerial habits, cost effectiveness and breeding performance. The most striking contrast is between climates that allow horses to be bred and reared in the open and those that require them to be stabled for long periods. Housing horses entails the use of people to lead them to and from pasture, individual feeding, the provision of bedding and its mucking out each day. These impositions are expensive in terms of labour, materials and fodder, incurring high capital and running costs. Confinement to buildings also has the disadvantage of exposing horses to unnaturally dusty atmospheres, accentuated by bedding material such as straw and hay, both of which carry quantities of fungal spores which, when inhaled, challenge the horse's immune system and lead to conditions such as 'broken wind' or chronic coughing. The risk of a dusty atmosphere may be compounded by closing stable doors to conserve heat loss; although this manoeuvre may succeed in raising the temperature of the loose box, it further decreases the efficiency of ventilation which is usually insufficient in many traditional-type stalls and loose boxes. A further sequel to confinement is the restriction imposed on the amount of movement and exercise made by the individual in the period of restraint. Outside, the individual moves frequently in grazing and this helps to maintain body temperature, as heat is liberated in muscular activity. In the loose box, a horse cannot keep itself warm by exercise, and the heat loss by radiation is considerable. The roof and walls of most loose boxes (tiled, wooden or brick) have poor insulating properties; in a cold climate a great deal of body heat is lost through this route, apart from that dissipated by convection (draughts), evaporation (from damp surfaces) and conduction (directly from the body to the floor). Lack of exercise may also have adverse effects on blood circulation and metabolic function, resulting in difficulties at foaling and impairment of the health of the unborn foal.

All these dangers are avoided in systems of management where the mare runs out at pasture day and night throughout the year. The main disadvantage in these circumstances is the increased exposure of the pasture to faecal and, therefore, parasite contamination. The burden of parasitic ova (eggs) deposited on the pasture is proportional to the number of grazing hours per acre; the more horses pastured, and the longer they are defecating on the pasture, the greater the problem. In the Southern Hemisphere the size of most studfarms is such that the number of horses grazed per acre is considerably less than under more intensive systems in the Northern Hemisphere. Parasite control measures (see page 329 are nowadays practised in all countries, helping to reduce the risk of excessive parasite infestation.

Open-air systems allow for much greater freedom and a more natural herd existence in which social bonds between members of the group are less seriously interrupted than where horses are housed. Management in free-ranging systems, uncomplicated by climate, is able to control the herd with a minimal number of attendants. In Australia, I visited the Sedgenhoe Stud near Scone in New South Wales. This famous studfarm, belonging to Mr Lionel Israel, had 67 resident mares in 1979, only 5 of which were barren, and the number of mares increased in the breeding season to 200. But only four men were required to perform the duties of feeding, teasing and supervising mating, parasite and other veterinary control measures. One would expect on a European studfarm to find at least fifteen personnel in these circumstances.

Of course, dry, hot conditions bring problems not experienced in wetter, cooler climates – for example, the epidemics of summer pneumonia experienced on studfarms in Australia due to dust on pastures, particularly where mares and foals congregate in sheltering under trees against the sun. But whatever the difficulties, the results of live foal production are equivalent to those encountered in the UK. At Mr Ken Cox's Stockwell Stud, near Melbourne, Victoria, the imported stallion Showdown had the following statistics between the years 1966 and 1974 inclusive: 549 mares, of which 452 (82.3 per cent) conceived according to early pregnancy testing. Of these, 15.5 per cent slipped a foal or had a stillborn delivery, leaving 382 (69.6 per cent) live foals as the eventual produce, an average of 42.5 live foals per annum. The number of mares served during each year in the period ranged from 50 to 70 and the average number of services per mare from 1.45 to 1.86. These figures would compare favourably with any Thoroughbred studfarm in Europe or elsewhere.

Pasture Area

Freedom to move over many acres is a prerequisite of herd life for reasons, already explained, of reducing parasitic dangers and enabling members of the herd to have access to adequate nourishment. Under intensive systems of management, and where available pastures for horses are limited, extra fodder may be provided and worm control measures implemented. In Southern Hemisphere systems, pasture acreage is often greater than that available in Northern Hemisphere countries. At Bhima Stud in Hunter Valley, New South Wales, where three stallions stand, the farm covers 1,700 acres, 500 of which are irrigated. There can be few, if any, studfarms of comparable size in Europe. Irrigation forms a prominent part of studfarm management in parts of the world where the climate is dry; and fodder is often based on the growth of lucerne or other plants with deep roots, from which several crops may be harvested. Elevation is a substantial influence on rainfall and pasture growth. At Hagley Stud, owned by the late Dr Peter Irwin, in Southern Australia the rainfall is 32 inches per annum, but reduces at the rate of about four inches per mile as land slopes away from the farm to lower lying areas.

Fencing

Fencing, whatever the area of grazing land, poses problems of economics and risk of injury. Barbed and plain-wire fencing is used commonly throughout the world to enclose grazing land used for cattle and, to a much lesser extent, pastures for horses. In the Southern Hemisphere, barbed wire is quite commonly used, even on equine studfarms, and the frequency of skin wounds is in consequence rather high. Wounds vary from small abrasions to deeply penetrating cuts involving the limbs, brisket and flank. Horses may be scarred for life. However, individuals that are reared in contact with barbed wire may learn to live with the hazard and be less at risk than individuals which are brought from one system of management to another. A young mare placed in a paddock with wire fencing, having been reared in conditions of post and rail fencing, may be more susceptible to injuring herself than if she had been exposed to this particular hazard from an early age. Wire netting of varying strength and link size has been found to be quite effective and relatively safe. Electrified wire fencing is used successfully in a number of studfarms in Australia. English-type layouts are also to be found on some studfarms in Australia. At Shirley Park, near Melbourne, there is no shortage of water and the paddocks are green throughout the year, fenced by white-painted wooden posts and rails to match the white and red-roofed buildings reminiscent of many English studfarms.

Parasite Control

Parasite control programmes (see page 329) do not vary greatly the world over. Drugs fed regularly, to reduce the amount of eggs deposited in the faeces on the pastures, is common practice, together with periods of rest for the pasture and concurrent or alternate grazing by cattle or sheep. Horses are notoriously destructive to pasture due to their grazing and eliminative behaviour which leaves some areas of overgrowth and others closely cropped. These latter areas tend to re-form even after ploughing and reseeding because they are contaminated, almost permanently, by equine excreta. Grazing by other species helps to reduce, but does not completely abolish, the imbalance. Similarly, periods of rest from horses and repeated mowing of pasture may go some way to diminishing the problem of deteriorating herbage.

Teasing

A wide variety of methods are used for teasing mares. The basic method, practised throughout the world, is to bring the mare into contact with the male for short periods every day or every other day (see page 77). This subjects her to intense sexual stimulation and, although unnatural, induces her to show signs of her sexual state that can be interpreted by those present; a decision is then made as to whether or not the mare should be mated. Varying emphasis is given in different situations. In general, Southern Hemisphere management employs restraint in a crush or stocks, through which the mares are herded one at a time, the teaser being placed outside the crush. This contrasts with the European custom of leading mares in hand to the teasing board (see page 77). The practical difference between these approaches is minimal. The determining factors are the length of time for which the mare is in contact with the male, and the degree to which her natural instincts to show signs in these circumstances are overridden by extraneous influences. These may be fear of unaccustomed surroundings, concern at being separated from her foal and other members of the herd, or a desire to be free to gallop or feed. It is important to recognise that the process of teasing and its objective may be defeated if account is not taken of environmental situations which mask the sexual state at any particular time.

It is perhaps more important to observe mares under situations of least stress, ie when they are grazing in the paddock together with other members of the herd. In the Southern Hemisphere, management favours observing mares at pasture in constant or frequent contact with the male. This contact may be achieved by confining a pony stallion in a railed or semi-boarded area surrounded by paddocks, so that mares in heat may at once show interest in the stallion. Riding or leading a pony stallion through the paddocks is effective if mares and stallions are accustomed to this practice. However, it can be dangerous to the rider, horse and mares if excitable or unmanageable stallions are used. In Europe it is more customary to lead the stallion beside the paddocks so that the handler observes the mares' response to the approach of the male horse.

Southern Hemisphere practice allows foals at foot to accompany their dams to the teasing board and to be present in a pen within the covering shed at the time of mating. This helps to diminish the anxiety of some foal-proud mares that do not conform to expected behavioural patterns when separated from their offspring.

Veterinary Examinations

Veterinary control of mating programmes, by pre-service examinations, is extensively practised throughout the world. The techniques employed do not vary to any great degree. They are based on rectal palpation and vaginal inspection, together with tests for bacterial infections of the genital tract. The most conspicuous difference concerns the method of handling; in the UK and Ireland mares are usually examined at the entrance to the stall or loose box, whereas in the Southern Hemisphere stocks are favoured. In Australia it is quite common practice for a hundred mares or more to be examined for pregnancy diagnosis by a veterinarian on one day. This number would be more unusual in stud farms in the UK.

Mating

The covering yard or mating shed is universally a simple area where mare and stallion meet, often for the first time. There may be stocks in which the mare may be examined or washed down before coitus, and in some studfarms the stallion too is washed before and/or after service. There may be a pen, should the mare have a foal at foot, into which the foal is placed, although this is more common practice in the Southern than in the Northern Hemisphere. Most covering yards have a teasing board which is permanent or can be swung into position for a final teasing of the mare before she is actually presented to the stallion. Restraint of the mare is frequently practised by means of bridle and twitch, and a foreleg strapped in a flexed position until the stallion has mounted. Felt boots are placed on the mare's hind feet and hobbles may be applied, although this latter practice has been largely discarded in the UK, where it is considered unnecessary and dangerous.

Foaling

Foaling in the Northern Hemisphere takes place almost always in a loose box. Over ninety per cent of mares foal between 7pm and 6am. This pattern is experienced in the Southern Hemisphere, where mares foal out of doors. Observation, in these conditions, is usually maintained by a nightwatchman who sits or sleeps in a caravan or a hut situated next to the foaling paddock, which may be lit by quartz iodine or other types of artificial lighting. There may be problems when mares foal in paddocks, with other members of the herd. For example, a mare that has not yet foaled may attempt to isolate a newborn foal from its mother. In other instances, a mare who has already foaled may poach the newly delivered foal and adopt it at the expense of her own foal. Some individuals, usually those foaling for the first time, gallop away from their foals, frightened by the afterbirth hanging behind them. It is a curious fact that even mares foaling in large paddocks often foal close to the perimeter fence; and it is not unknown for a mare to deliver her foal in such circumstances that it falls down a bank and into a pond with fatal consequences. These problems are not, of course, encountered in the confines of a foaling box in the Northern Hemisphere managerial systems.

The newborn foal's navel is treated with iodine or antiseptic; enemas are administered and assistance given to the foal to help it to suck from the mare if necessary. Foals may variously be given orange juice, glycerine, epsom salts, castor oil and/or white of egg as soon as possible after birth in an attempt to prevent or alleviate problems of passing the first dung, ie meconium (see page 216).

Dates of Breeding Season

Wherever one resides in the world of horse breeding, two hotly contested debates can be heard: the need to move the arbitrarily selected dates of the Thoroughbred breeding season to coincide with the natural breeding season (see page 37), and the use of artificial insemination (AI) (see page 46). Most vets agree that it would be helpful to change the dates of the breeding season to obtain better fertility and to ensure that foals are born during optimal climatic conditions and pasture growth. However, the arguments for and against change are greatly influenced by such factors as economics and the particular climate of the area – some being more extreme than others. The real test is whether or not altering the dates would bring benefits in, firstly, reducing overheads and, secondly, increasing productivity. In the UK and Ireland, to move the starting dates of the Thoroughbred breeding season from the present 15 February to 1 April would undoubtedly reduce overheads and probably increase productivity. Under the present system there is a great deal of effort required to overcome the mare's inherent tendency to be sexually inactive, ie anoestrus (see page 74), before April; to correct the imbalance, management has to invest substantial sums in terms of stabling, artificial lighting, fodder, etc – quite apart from the costs of veterinary expertise and treatment which breeding in the winter months entails. A further dimension in the argument is that services in winter and early spring result in foals being born in winter months, when the need for protective measures is correspondingly increased. Another aspect of the problem is that the number of nights on which 'sitting up' is required is directly proportional to the length of the breeding season. Conversely, to concentrate the period in which mating occurs correspondingly decreases the labour costs involved, and diminishes the expense of boarding charges incurred by owners who have their own establishments and can thereby maintain their mares at less cost at home. These factors do not necessarily apply in climates where mares can be kept outdoors day and night throughout the year. Nor, in these circumstances, is winter and early spring anoestrus (inactivity) as marked as where winter and summer climates are extreme. However, anoestrus is largely related to duration of daylight hours, and the geographical latitude may therefore play a significant part in determining the difficulty or otherwise of breeding outside the natural months.

Artificial Insemination (AI)

AI is practised in certain countries and on certain breeds, eg the Standardbred horses of the USA and Arabian, Hanoverian and many other breeds in countries such as Germany, China, Poland and the USSR. There is one breed, the Thoroughbred, whose authorities adamantly oppose the use of the technique. There has been recent strong pressure from veterinary groups to persuade the Thoroughbred stud book authorities to retreat from their attitude of comprehensive prohibition and to allow AI to be used strictly for veterinary reasons. These reasons include the need to avoid venereal infection where mares are known to be, or to have been, infected with a venereal microbe such as *Klebsiella* or the organism of CEM. The various techniques of AI are described in chapter 15 and the arguments for and against its use on Thoroughbreds for veterinary reasons are discussed on page 131. Here we are concerned only with the question of AI relative to improved productivity. There can be little doubt that AI, as used by Standardbred and trotting breeds in the USA, decreases the amount of labour required for planned natural matings, increases the likelihood of conception, and enables greater numbers of mares to be bred and to conceive than would otherwise be the case. In other words, AI enables studfarms to produce more foals at less cost than they could achieve if natural coitus were employed. Unfortunately these arguments are rejected by Thoroughbred authorities, who fear that the breed may be harmed by the over-use of certain bloodlines at the expense of others. There is in addition a traditional objection to AI typified by the declaration of the great Italian breeder Signor Tesio, who opposed AI on the grounds that it removed the element of 'love' and that no classic winner had ever been conceived in artificial circumstances. Nowadays we can hardly subscribe to this supposition, but one can sympathise with the argument that splitting ejaculates, the basic advantage of AI, might have a deleterious effect on the financial interests of breeders whose solvency may depend on scarcity, rather than on abundance, of certain lines within the breed.

Another argument, not always considered when these matters are discussed, is the relatively poor semen quality of Thoroughbred stallions. In chapter 4, we shall encounter the fact that conception depends on a minimum number of live normal sperm being deposited in the uterus of the mare. In many breeds where AI is routinely employed stallions are selected for the quality of their semen, but in Thoroughbreds this is not the case and stallions are selected almost entirely on their racecourse performance. Many Thoroughbred stallions have rather low numbers of live normal sperm in their ejaculates, and to split these samples may reduce the numbers in each sample to subminimum proportions, thus depriving the inseminated mare of the chance of conception that she would otherwise enjoy had the insemination been natural and contained the whole ejaculate.

AI is, therefore, a useful technique for preventing the spread of venereal disease and helpful to management and commercial interests in certain well-defined circumstances; but it cannot be expected to contribute to productivity in breeds such as the Thoroughbred unless there is a radical departure from existing attitudes.

PRODUCTIVITY

Productivity can be measured in a number of ways. For example, breeders may take a financial endpoint related to success in selling produce and/or winning prizes. The basis for this assessment rests largely on market forces, fashion and the subjective judgement of purchasers.

The most useful indicators of productivity are those relating to efficiency, such as the percentage of live foals and their soundness for the purpose for which they are bred. The Thoroughbred Stud Book returns, because they are comprehensive, serve to illustrate the level of productivity in this particular breed, and may be applicable to other breeds. A survey carried out by BEVA showed that for every 100 mares for which returns are made, of those covered by Thoroughbred stallions, 25 are barren, and 10 abort or have dead foals. This leaves 65 which may be expected to be available for training. However, of these only 28 actually run in races at any age. The remainder are exported, considered too inferior to race, and/or become ill or unsound. It may be concluded that it takes 100 mares to produce 28 living, sound produce each year or, in other words, about 10,000 mares to produce a crop of 3,400 two-year-olds, which is the approximate number in training each year on the flat. These figures, taken from the whole population, do not accurately reflect localised situations where special factors of excellence in terms of selection, management and husbandry enable significantly better results to be obtained on small samples of the population.

A similar survey to that conducted by BEVA has been reported by Dr J. M. Bourke of the Victoria Racing Club, Australia, who listed some of the reasons for wastage among the 2,000 new horses which were introduced into the racehorse population each year in Victoria. Dr Bourke listed the reasons for wastage as export from the state or county, failure to stand training, premature retirement through lack of ability, injury, or some other factor. It is clear that although wastage is not entirely due to biological processes there are substantial areas in which veterinary science and/or management might play a beneficial role.

It is often argued that there are sufficient horses available for training and that we should do nothing to increase the already swollen population of Thoroughbred horses in the country. However, an increase in productivity could be used to effect a proportional decrease in the total population of mares; with a consequent annual saving of current expenditure and a further reduction in capital investment. For example a ten-per-cent reduction in the barren and live foal wastage in any one year represents only about four per cent of the total overall loss. So there seem to be abundant opportunities for much more substantial savings at all levels if a concerted effort were made by breeders and vets, employing veterinary science and common sense. This is the challenge facing all sections of the horse breeding community.

Culling for Positive Reasons

The raising of productivity, in percentage terms, would enable breeders to select stock of inferior quality for culling, thus raising the standard of the breed. The key to progress lies in identifying the areas where progress could be made. There is for certain no panacea or single measure by which the annual wastage could be significantly reduced. Further, any measure that might create a marked difference would need the support of all concerned – vets, owners, managers and studmen. From a veterinary viewpoint, I suggest that more attention should be paid to selecting for positive values in brood mares, such as the ability to conceive and deliver a live, healthy foal of normal birth weight every year. Mares falling below a given standard in this respect should be culled. A similar approach to stallions might be helpful, but breeders would obviously be more reluctant to cull males than females. For example, a Derby winner is a commercial proposition at stud, even though he obtains only forty to fifty per cent conception rates among his mares.

The relationship between racecourse performance and selection for stud is much closer in stallions than in mares. In practice, many mares are retired to stud duties because they cannot win races or are only able to scrape home in a three-year-old maiden race on a minor racecourse. The chances are, therefore, that mares may be selected more readily for breeding as opposed to racing performance. It cannot be overemphasised that the influence of the dam on the physical make-up of the progeny is greater than that of the sire, because the mare has not only to conceive but to carry, deliver and nourish the foal from conception to weaning. The heritability of racecourse performance has been shown to be approximately 33 per cent in a study conducted by the Department of Genetics, Trinity College, Dublin. This means that the racing ability of any individual is due to approximately 33 per cent genetic effects and 66 per cent environmental effects. Genetic effects are the inherited components at the moment of conception, ie the genetic 'blueprint', and environmental effects exert their influence from conception through both intrauterine and extrauterine growth.

These findings highlight the importance of environmental effects, which are the day-to-day concern of management and the veterinary surgeon; but once the environment is relatively standardised, or optimal, improvement in progeny can only be achieved through improvement in the inherited components.

Fifteen per cent of all the colts that race in the UK and Ireland are selected for use as stallions, whereas eighty-five per cent of fillies that race are retained for breeding. The selection pressure on stallions is therefore so great as to limit the range in genetic merit between them. There is a much greater range in genetic merit between dams, and so improvement in the genetic merit of offspring is more easily achieved by careful selection of the dam than by sire selection.

Co-operation with Veterinary Science

A more thoughtful use of veterinary science should be encouraged. For example, all barren mares should be examined within four months of the end of the breeding season. It is important to assess the reason for their barrenness and, following this diagnosis, to apply appropriate treatment to provide an improved chance of conception in the subsequent season. Mares nowadays often arrive on studfarms to be mated suffering from some genital condition requiring treatment that should have been instituted before, rather than during, the period of mating (see Infertility, page 245).

Owners do not always co-operate with vets to their best interests in managing healthy mares. For instance, the veterinary examinations that are such a feature of the breeding season bring advantages discussed on page 238, but many owners do not take advantage of out-of-breeding season examinations. Mares are often sent to the studfarm in February, and it is assumed by their owners that they will be both ready for mating early in the breeding season, and in a suitably fertile condition. However, many individuals are not in a sexually active state and could be retained at home for some weeks or months before being dispatched to incur the added expense of an 'away' studfarm. All that is required is to have the mare examined from the start of the stud season by a vet at, say, fortnightly intervals until ovarian activity is found. A check of the blood progesterone level should be made to ascertain whether or not lack of sexual activity is due to the presence of a yellow body in the ovary (for an explanation of this reasoning see page 73). An owner may take special measures to induce sexual activity in the winter by applying artificial lighting to extend daylight hours (see page 37). Veterinary examinations will, however, still be necessary to assess whether or not these measures have been successful.

An efficient breeding programme must contain suitable measures to prevent loss through parasite infestation, bacterial and viral infection, nutritional disorders, and injuries. It is essential for breeders and vets to co-operate in maintaining constant vigilance to ensure that preventive measures are adequate and continuing, if serious consequences are to be avoided. For example, the parasite control programme must be tested for effectiveness at least once a year and, if found wanting, the programme should be altered accordingly (see page 329 for discussion on parasite control). Similarly, vaccination programmes against such diseases as influenza and tetanus (lockjaw) must be implemented in such a way that they protect at all times – in particular those members of the herd, such as foals, which are most at risk to the effects of these diseases.

Inadequate nutrition or a diet lacking a proper balance of vitamins, minerals, proteins and carbohydrates may result in poor growth, bone disorders and unthriftiness. Failure to follow the principles of good husbandry may lead to various impairments of growth and development in the foetal foal or, even, to abortion. There are many gaps in our knowledge of the horse's nutritional needs, but breeders are advised to study this cause of possible wastage in the light of current information on the subject, in conjunction with the advice proffered by nutritional scientists and veterinarians. It is important for management to reassess their procedures annually on the basis of the results they obtain in all areas of productivity – percentage output, and the health, growth and soundness of their produce. Managerial malpractices, such as overcrowding mares, foals and/or yearlings on premises with limited buildings and pasture facilities, poor pasture management, feeding of inferior foodstuff and failure to adhere to sound scientific principles of disease prevention and control, can be responsible for serious economic loss; the result is poor productivity.

The Myth and Reality

Horse breeding may be likened to a multicoloured prism with many facets, changing its appearance according to the angle from which it is viewed. It is an enterprise offering rewards of considerable magnitude for those who seek money, achievement or merely the considerable pleasure that can be gained from participating in the miracle of procreation as it applies to one of nature's most endearing creatures. The in-foal mare kept on rough land behind the owner's bungalow, with access to a ramshackle shed for shelter, may supply all these needs – profit, pleasure and a sense of achievement – just as much as one that is maintained with numerous others on an establishment possessing immaculate buildings, vast acreages of well-fenced land, managed by professional staff. In practice, man's aspiration often determines the managerial environment, the crossbred pony of low value being tended by its owner with human care, whereas the valuable Thoroughbred is the subject of intensive management. But whether pony, Thoroughbred or any other type, a horse is basically a horse, and responds to the challenges of its environment as a horse, not according to human requirements. The adage 'you can lead a horse to water but you can't make it drink' is most appropriate in the context of management and biological function. Horses are very adaptable – they can breed in desertlike conditions and in swamps, in low-lying pastures or in mountain scrubland – but there are limits to their versatility; and it is these limits to which we are answerable in our quest for breeding horses selectively, economically and for specific purposes. When man lays down the conditions of breed, type and performance as the chief objective of mating, biological adaptive responses tend to be ignored.

Let us take an example of how selection for an artificial objective may interfere with the process of natural selection. A group of Thoroughbred mares and stallions have been breeding for generations in a particular environment in which the stallions are housed but the mares run free day and night; the climate is very mild throughout the year. In selecting the mares for breeding, little note is taken of their performance on the race-course, but the stallions are used only if they have outstanding racing ability. Suppose that a group of these mares is moved to another country where there are extremes of climate and the mares now have to be confined to loose boxes for long periods. This environment would present a new challenge; some mares would adapt (become acclimatised) but some individuals might not. If we were then to select the mares in the new environment on the basis of racing performance we would introduce a further influence affecting the adaptive responses of their progeny. After a few generations within the new environment, we would expect to find resident (indigenous) stock fully adapted to this environment, and recent 'immigrants' of which some were able to adapt while others had problems ranging to complete inability to do so. Failure might be manifested by such signs as behavioural traits (vices, bad habits), unthriftiness, suscepti- bility to endemic diseases, and irregularities in reproductive function. Some mares might have perverse sexual activity, being inactive during months included in the arbitrarily defined breeding season and active in autumn and winter.

In practice, if we examine any given population of mares, we find a proportion of individuals having a biological make-up that is at variance with the environmental challenge, thus subjecting these individuals to stressful circumstances. The individual that thrives while in the paddock but loses condition in training is probably an example of failure in the adaptive response to confinement and to isolation from other members of the herd. It is not possible to identify the exact relationship between the effects of selection, environment and the ability to adapt to any particular environmental circumstances, but we can be sure that these influences are at work within our mare and stallion populations; and we should take note of this possibility when we manipulate the environment for our own ends. We may expect about 50 per cent of the population to conform, 25 per cent to acquiesce reluctantly, and 25 per cent, in one way or another, to disappoint us. These are the biological facts of life; the only way to improve our results in practice is to understand the fundamental aspects of how the body functions for breeding and how malfunction operates in terms of infertility and diseases. These interesting subjects form the matter of Parts II and III of this book; they represent the alpha and the omega of success and failure in the breeding of horses, with whichever breed or type we are concerned.

Education of Studfarm Personnel

Education and experience are two sides of the same coin. People who look after horses, and particularly those on studfarms and smaller breeding establishments, have always relied on experience and learning by example, rather than on a formal education. Knowledge and experience have been handed down from generation to generation of studmen, the studgroom forming the linchpin of the apprentice system. There have been few attempts to place education on an organised and rational footing. The Thoroughbred breeding industry has preferred to stand aside or to pay only nominal lip-service to any concerted effort to educate its working personnel. This is despite the huge capital investment in horses and the fact that a stallion man may be responsible for the care of a horse worth millions of pounds and a studgroom be prepared to deliver a foal worth thousands of pounds from a mare valued in five figures. The reasons for this reluctance are chiefly that it has suited owners to maintain traditional hierarchical systems of employment in which they are not required to remunerate studmen at rates above the level of agricultural workers, at the same time avoiding the development of a career structure with a high degree of professionalism. I do not wish to suggest here that this has been a deliberate attitude, but rather a matter of *laissez-faire*. Owners do not wish to exploit stud personnel but often they do not appreciate fully the extent to which success or failure depends on the integrity and skill of their working teams.

There have been some notable exceptions among organisations and individuals. For example, the Thoroughbred Breeders' Association has provided lecture courses and seminars for studmen, stud managers, and even owners. However, valuable as these have been, the demand has outweighed the supply, and in any case efforts have been fragmentary and confined in the most part to annual one- or two-day events. The Irish National Stud has achieved much by its courses of instruction, which were started in 1970. These training courses in horse breeding are for men and women aged eighteen to twenty-five years and last for six months. They are residential and intended for studgrooms and stud managers. Practical work includes the usual stable routine, and attendance at foalings. Lectures are provided on feeding, housing, grooming and control of the horse, the management of grassland, worm control, accounting systems, disease conditions, breeding, farriery and other aspects. The course is supervised by Michael Osborne, a veterinarian in charge of the National Stud. The number of people attending the course has ranged from ten to twenty-two; written, oral and practical examinations are held at the end of the course, when a certificate of merit is awarded to those who reach a given standard. The English National Stud, under its director Miles Littlewort, has more recently introduced courses of a similar nature for students working on the studfarm during the breeding season. In Ireland, a number of programmes involving research into management and husbandry have also been implemented. There can be little doubt that training and research should go hand in hand; the development of new methods through original observations is as integral a part of the development as it is of veterinary science.

An interesting experiment was started at the Derisley Wood Stud in Newmarket by the late Mr Irvine Allen. This philanthropic owner, a governor of one of England's most progressive public schools, Millfield, founded the Newmarket School of Stud Management in 1970. In the first year the Meddler, Stetchworth Park and National studs collaborated in running a nine-month course for people interested in furthering their knowledge of studfarm procedures. The scheme failed to gain support and foundered on the rocks of disagreement on such vital matters as who should be accepted (working studmen, managers, bloodstock agents and/or those who required a suitable haven for marking time before deciding on their future career), who would employ the graduates of such a course, and whether they would be the educated upstarts in a world of practical people. These were familiar objections, carrying some degree of truth but ignoring the fact that the industry is in great need of a system of education of one kind or another. The ideal can only be evolved by collaboration and some degree of trial and error. A further argument against the Newmarket course was the fact that the teaching was too highly orientated towards veterinary instruction. This argument did indeed have some validity; nonetheless, it seems appropriate that vets should lead the way. They have a professional training and are therefore more knowledgeable and able to provide instruction. The teaching of stud management procedures by stud personnel demands, over a period of time, the development of teachers who have themselves previously been taught – a chicken and egg situation. The most disappointing aspect of the failure of this experiment was that it had the tacit support of the Horserace Betting Levy Board and the National Stud, both of which organisations, one would have thought, might take the initiative in such a venture. It is essential, in my opinion, to institute a career structure for stud personnel, rising from ordinary studmen to stallion men, studgrooms and stud managers in a progression of knowledge, experience and responsibility, all trained to increasing standards of efficiency. This should not be taken to suggest that present-day studmen, at any of these levels, are not highly competent, but writing as a vet I can appreciate the advantages which are bestowed by training and especially by a recognition of that training in terms of status, security of employment, incentive and reward.

THE REPRODUCTIVE CYCLE

Reproduction is a chain of biological events, starting with two individuals in one generation. These conceive a new individual, nourish it within the female uterus, from which it is expelled at birth, and further nourish it until it is weaned to independent existence. The new individual is thus, when it becomes sexually mature, capable of initiating a further sequence of reproduction, continuing the reproductive process.

Conception is the beginning of the new individual and occurs at the moment of fertilisation of the egg (female gamete) and the sperm (male gamete). The male and female functions, described in the simplest terms, produce and deliver the male and female gametes to a meeting place where fertilisation can occur; the egg is passive, the sperm active and penetrating. The site of union occurs in the upper part of the oviduct. The female has the added responsibility of nurturing the new individual throughout the period known as pregnancy. The female organs consist of two ovaries, producing eggs, and the genital tract, a tube along which the sperm can pass to the egg and in which the fertilised egg can develop until it is ready to be delivered at birth.

The male organs comprise the two testes (testicles), which produce the sperm, and the male genital tract, along which the sperm are propelled at the time of mating to be deposited in the female genital tract. All the organs and glands of the sexual system of the male and female are 'designed' and adapted for the events of fertilisation, gestation (pregnancy) and birth. Finally, the female mammary organs (udder) provide nourishment for the foal until it is weaned.

The purpose of Parts II and III of this book is to explore the background of the reproductive story, in health and disease.

PART II

NATURAL FUNCTIONS OF THE HORSE

3

THE SEXUAL FUNCTIONS
OF THE MARE

THE OESTROUS CYCLE

The oestrous cycle, as the term implies, is a series of events recurring at intervals. Oestrus (heat) lasts, typically, for five days, and dioestrus for fifteen days. Thus the typical oestrous cycle covers twenty days in all.

From a strictly biological viewpoint, oestrus is the period when the mare accepts the stallion; in stud parlance, she is said to be 'in heat', 'in season', 'on' or 'showing'. Dioestrus means, literally, in between oestrus. During dioestrus the mare will not accept the stallion and she is said to be 'off', 'out of heat', 'out of season' or 'not showing'. In practice, the oestrous cycle is regarded as a sequence of recurring periods of oestrus.

The egg (ovum) is shed from the ovary during oestrus and thereby becomes available for fertilisation by the stallion's sperm. Dioestrus, on the other hand, is an interval during which the uterus is prepared to receive the fertilised egg should conception occur. If the egg is not fertilised, this preparatory period for pregnancy is brought to an end and a further oestrous period promoted, providing another opportunity for mating and fertilisation. Once a fertilised egg has entered the uterus, oestrous cycles are suppressed and the individual maintains a state which is equivalent to that of dioestrus.

For the purpose of discussion, we can consider the oestrous cycle under the five headings of (1) anatomy, (2) physiology, (3) behaviour, (4) influences and variations, (5) knowledge applied in practice.

ANATOMY

The oestrous cycle is based upon the organs of reproduction and the glands that control their functions. The sex organs of the mare consist of the ovaries and the genital tract, which comprises the Fallopian tubes, uterus, cervix, vagina and vulva (fig 2). The uterus is capable of expanding to accommodate the developing foal for the eleven months of

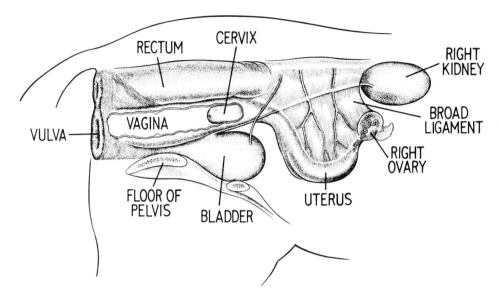

Fig 2 The genital organs of a mare seen from the right side; the ovary and uterus are suspended by the broad ligament containing blood vessels

pregnancy, and the cervix, vagina and vulva form the birth canal. The sex hormones are produced by the anterior pituitary gland, the ovaries and uterus. The reproductive organs include the mammary glands, but these do not form part of the oestrous cycle and are not of concern to us here.

The Ovaries

The ovaries are roughly bean-shaped, although their actual shape and size vary with the breed and age of the individual and the season of year. They measure on average $2\frac{1}{2}$in x $1\frac{1}{2}$in (7cm x 4cm) and each consists of a firm fibrous mass called the stroma. The individual is born with many thousands of eggs (ova) contained in the stroma just below the surface of the ovary; and no more are produced in the lifetime of the mare. Once a filly is sexually mature, at about two years old, a number of fluid sacs (follicles) develop around the eggs and increase in size during oestrus. The follicles act as an hormonal gland secreting oestrogen. Eventually one of the follicles ruptures (ovulates) and the egg escapes through a weak point in the capsule surrounding the ovary. This weak point is known as the ovulation fossa and is situated at the indented part of the 'bean'. The lining of the follicle bleeds, and a blood clot forms in the space previously occupied by the fluid and the egg. Special cells in the lining of the follicle, known as luteal cells, then grow into the clot and form an hormonal gland called a yellow body or corpus luteum. This gland secretes progesterone. The yellow body functions during dioestrus. In the winter months there is no activity in the ovaries, a condition known as anoestrus. There are neither follicles nor functional yellow bodies present in this state of sexual

Preserved specimens of four ovaries cut in half to show the various structures contained: (bottom left) two small follicles abutting one another can be seen in the lower half with a dark structure being a recently formed yellow body, immediately above. The lighter-coloured patch next to the yellow body is an old and non-functioning yellow body; (bottom right) two large recently formed yellow bodies can be seen in the upper half of this ovary with an old yellow body immediately below and a thick-walled small follicle beyond; (upper right) this ovary contains a large mature follicle and some scarring caused by yellow bodies which have long ceased to function; (top left) an unusual ovary containing multiple follicles of varying sizes

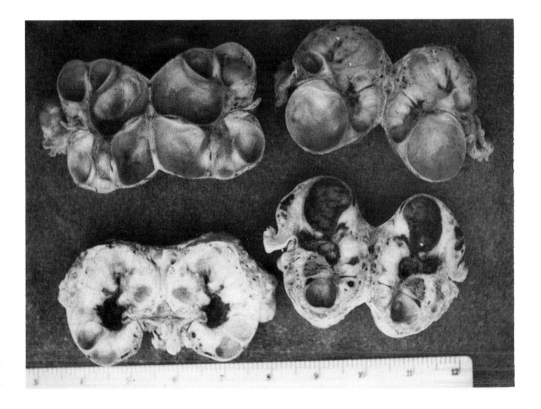

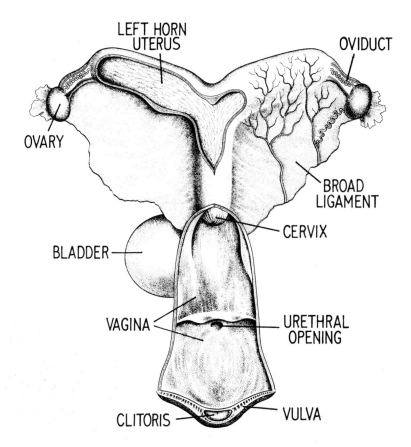

Fig 3 Ovaries, broad ligament, oviduct, uterus (partly opened), cervix and vagina of mare viewed from above; the bladder can be seen below the vagina and its opening (urethra) just in front of the fold representing the hymen two-thirds way down the vagina

quiescence. The ovaries, therefore, contain various structures depending on the sexual state of the individual at a particular time.

The Fallopian Tubes

The Fallopian tubes or oviducts lie in the membrane that supports the ovaries and uterus (fig 3). Each tube is coiled and measures about 10in (25cm). At the ovarian end, the tubes open directly into the uterus at an opening on the tip of each horn. The tubes are about 0.08-0.1in (2–3mm) in diameter at the uterine end, but near the ovary they widen considerably to form an ampulla. It is here that fertilisation occurs between the egg, which is shed into the tube at the time of ovulation, and the sperm, which make their way through the uterus and up the tubes after coitus.

The minute structure of the tube is seen as an ultra-fine outer lining, a middle muscular layer and an inner mucous layer. The mucous membrane lining contains a single layer of cells, with cilia (threads) producing a current directed towards the uterus. The function of the tubes is to receive the egg and sperm and to allow the fertilised egg to pass into the uterus. It is an odd fact, as yet not fully explained, that in the mare unfertilised eggs remain in the Fallopian tube and do not pass into the uterus.

A disposable cardboard speculum inserted into the vagina of a heavily stitched mare. The cervix may be observed, and swabs taken, aided by a pencil battery torch

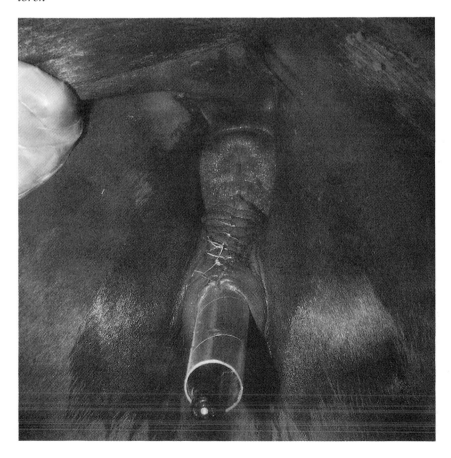

The Uterus

The uterus is a hollow muscular organ with a body and two horns giving it a roughly Y-shaped appearance (fig 3). The organ is suspended from the roof of the abdominal cavity by a membrane or mesentery known as the broad ligament. The arteries, veins and nerves which supply the uterus and ovaries run in this membrane. Rupture of the artery at the time of foaling may give rise to serious, and possibly fatal, haemorrhage (see page 293). In the non-pregnant state the uterus is relatively small compared with the pregnant and full-term size. The horns measure about 10in (25cm) and the body about 8in (20cm) in length. The body is about 4in (10cm) in diameter. However, the form and dimensions of the organ vary with the sexual state, as we will discuss later. At the hind part, the body of the uterus constricts to a neck or cervix which separates the organ from the vagina. The cervix is about 3in (7.5cm) long and 1½in (4cm) in diameter. Part of the cervix projects into the cavity of the vagina and can be seen through a speculum inserted into the vagina as a roselike structure. The cervix also varies considerably according to the sexual state of the mare.

The cavity of the uterus is largely obliterated in the non-pregnant mare and it communicates at the tip of each horn with the Fallopian tubes and at the opposite end with the cervix. The wall consists of three coats, an outer fine membrane continuous with the broad ligament, a middle muscular coat, and an inner mucous membrane lining. We will see later, when discussing biopsy of the uterine lining (endometrium), that the lining contains a single layer of epithelial cells below which are numerous branched tubular glands.

During oestrus the walls of the uterus are relaxed and flaccid, the uterine glands secrete moist mucus, and the uterine surface is engorged with blood. During dioestrus the muscle in the uterine wall has considerable tone, and the organ is turgid. The uterine glands are inactive, the mucus they secrete is sticky, and there is little blood flow in the uterine lining.

The uterus is capable of undergoing extensive changes during pregnancy, when it has to accommodate the developing foal, besides acting as a surface for attachment of the placenta.

The Cervix

The cervix relaxes or constricts according to the sexual state, being open in oestrus and closed in dioestrus (and pregnancy). It can be viewed through the vagina as red, moist and relaxed during oestrus – constricted, pale and sticky during dioestrus.

The Vagina

The vagina extends through the pelvic cavity from the neck of the uterus to the vulva. It is tubular and measures about 8in (20cm) long. In the normal, natural state its walls are apposed to one another but, if the vestibular seal (see page 67) is breached by a speculum or the horse's penis, the walls diverge and the diameter of the vagina proves to be about 5in (12cm). In fact, its ability to dilate is limited only by the bones of the surrounding pelvic girdle.

The wall of the vagina consists of a mucous inner coat and a muscular outer coat, both of which are highly elastic. The vaginal wall does not contain glands and only its very anterior part is covered by peritoneum. The vaginal cavity is divided into a forward (anterior) and a backward (posterior) part by a fold of the vaginal wall which, only rarely, forms a complete membrane or hymen. The urinary duct (urethra) opens just behind the region of the hymen in the centre of the posterior vaginal floor. It is in this opening that we may find microbes, such as *Klebsiella*, which can also be the cause of uterine infections (see page 239).

The Vulva

The vulva is the most posterior part of the genital tract. It guards the entrance to the vagina although there is no clear line of demarcation between the two. The posterior vagina may therefore be considered as part of the vulva, although the structure that we usually refer to as the vulva is that of the external orifice formed by the vulval labia or lips. These meet at an angle above and below called the dorsal and ventral commissures.

The vulval lips are covered by thin, pigmented, smooth skin richly supplied with sebaceous and sweat glands. Under the skin there is a layer of muscle capable of constricting the vulva, which fuses above with the sphincter of the anus and, below it, surrounds the clitoris. This muscle is partly responsible for the shortening and lengthening of the vulva in, respectively, dioestrus and oestrus. Together with another muscle that surrounds the clitoris it is capable of opening the vulval lips to expose the clitoris in the action known, colloquially, as 'winking'.

The Clitoris

When the lips of the vulva are drawn apart it is possible to observe a rounded body about an inch (2.5cm) wide occupying a cavity in the ventral commissure. This is the glans clitoridis, the equivalent of the glans penis of the male. The cavity in which the glans lies is the fossa clitoridis. The roof of this cavity or fossa is formed by a thin fold or frenulum. In the centre of the glans is a sinus which contains a fatty exudate or smegma similar to that found in the prepuce of the stallion. It is these materials, the sinus and the clitoral fossa, which have received considerable attention in recent years due to the fact that they may harbour venereal microbes such as that causing contagious equine metritis or other infections.

The clitoris exposed by parting the lower end of the vulva: it is separated from the vagina by a membrane (fraenulum, arrowed), and consists of a body (centre) surrounded by a fossa. The sinus is situated in the middle of the body

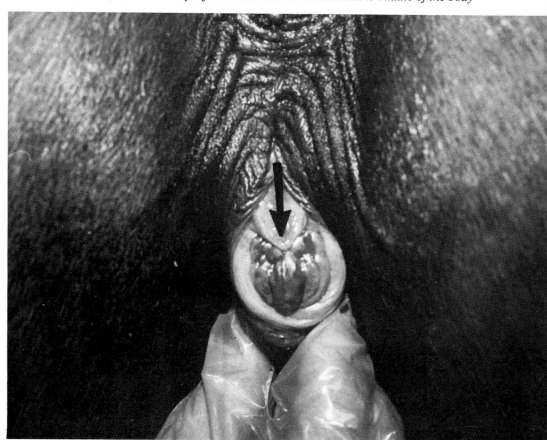

The Conformation of the Posterior Genital Tract

For many years it has been recognised that air may, abnormally, enter the genital tract, causing infection and infertility. This subject is discussed more fully later (see page 249). The normal conformation of the vagina and the vulva, relative to the bony floor formed by the pelvis, is shown in fig 4. Three seals prevent air being sucked into the genital tract by the potential vacuum present in the uterus and vagina. These seals are (1) the vulval lips, (2) the vestibular seal formed by the fold in the wall of the vagina at the level of the hymen and the pelvic brim, (3) the cervix. The seals are affected by oestrus when the vulval lips are lengthened and slack, the vestibular seal moistened by mucus and the cervix relaxed.

The vulva is contracted, the mucus sticky and the cervix tightly closed during dioestrus, and the seals are therefore less likely to be breached, except in cases of poor conformation (fig 5). Air may also enter the vagina when a speculum is introduced, during coitus and following birth when the vagina and vulva are greatly distended.

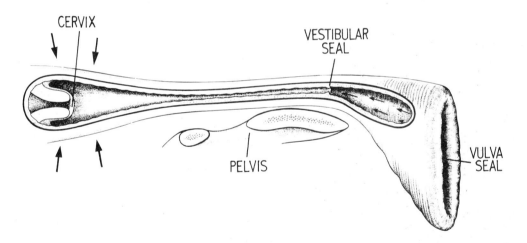

Fig 4 Normal conformation of mare's vagina illustrating seals preventing air being sucked into tract (after R. M. Butterfield)

PHYSIOLOGY

Cyclic Changes in the Breeding Organs of the Mare

Heat Changes in the genital organs correspond to those of sexual behaviour. During oestrus, the genital tract is relaxed and its surface

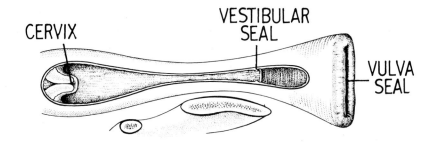

CERVIX   VESTIBULAR SEAL   VULVA SEAL

OFF HEAT — If vulva seal effective –
cervix closed. Vestibular
seal not needed.

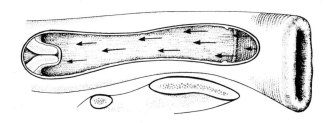

OFF HEAT — Vulva and vestibular seal
not effective.

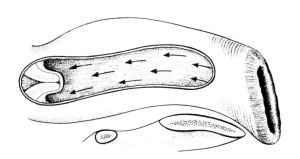

DITTO — Old or poor mare – same only worse.

Fig 5 Poor conformation mare genitalia

lubricated by a moist mucous covering. The vulval lips are swollen and lengthened, their surface moistened by mucus from the vagina. Muscles in the vulva enable the mare to relax or contract its length. The vulval lips may be everted ('winking') to expose the clitoris. This may be a self-stimulating process exposing this sensitive area, but it also has an important visual role in attracting the attention of the male.

If we open the vulval lips and look through a speculum into the vagina we can see that the lining of the vagina and cervix is moist and reddened and that the folds of the cervix are oedematous (swollen) and relaxed, enlarging the opening into the uterus. We cannot normally see directly into the uterus except by using special optic instruments, but if we could we would be able to observe that during oestrus the uterine lining is puffy, engorged with blood and covered by a thin, moist mucous layer. If we were to take a microscopic view we would see that the cell lining of the uterine surface contained tall (columnar) epithelial cells and that in the deeper layers there were many active glands secreting mucus.

The blood-engorged, relaxed and surface-moistened state of the genital tract is suited to the entry of the stallion's penis and to the passage of the spermatozoa through the uterus to the Fallopian tubes. One or both of the ovaries, during oestrus, contain a number of small follicles (fig 6) according to individual and seasonal variations. Their sizes range from a few millimetres to several centimetres in diameter. They are present just below the surface of the ovary and the larger ones may be felt per rectum at the time of a veterinary examination (see page 81) as soft to tense fluctuating swellings projecting from the ovarian surface. One (in some cases, two) of the follicles develops to a larger and more mature state than the others. This follicle(s) is destined to ovulate, that is, to rupture and shed the egg it contains. After the follicle has ruptured it is replaced by the yellow body, and within twenty-four to forty-eight hours the mare goes 'out of oestrus' and into a state of dioestrus.

In summary, oestrus is a period in which the egg is shed, coinciding with the behaviour of acceptance. The egg passes, together with fluid and debris from the ruptured follicle, into the Fallopian tube, where it is available for fertilisation by the stallion's sperm. The sperm, ideally, should be present in the Fallopian tube before ovulation.

Out of Heat Dioestrus is a state of preparedness for the development of the fertilised egg, which arrives in the uterus on the fifth or sixth day following conception. The uterine lining has by this time changed from a moist to a sticky dry condition. If the egg is not fertilised the genital tract

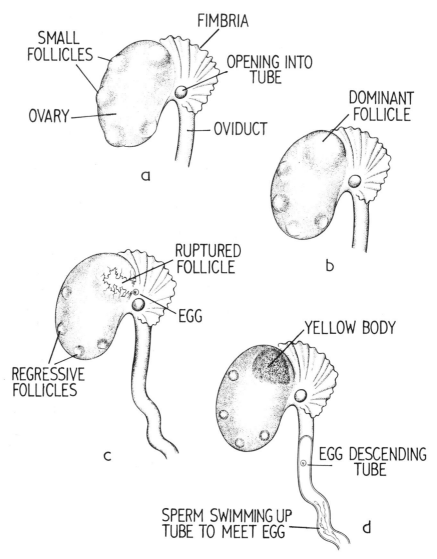

Fig 6 Events occurring in the ovary during oestrus

still undergoes this period of preparation, but it is cut short on about the fifteenth day to give rise to a further oestrous period, thus initiating the start of a new cycle.

In dioestrus the vulval lips are shortened and dry, the vaginal and cervical surfaces pale and covered by sticky mucus. The cervix is tightly shut. The uterus is firm or turgid, the cells of its surface lining (epithelium) reduced to a cuboidal shape and the underlying glands relatively inactive and small. The ovaries contain the yellow body of the previous oestrus. There are few, if any, follicles present, except that (1) smaller unruptured follicles of the previous oestrus may take time to diminish in size or regress, and (2) new follicles may be developing in mid dioestrus in preparation for the next heat period.

Natural Control of the Oestrous Cycle

The body controls its various functions by communication through the nerves and by hormones. Reproductive function is mainly under hormonal and, to a lesser but important extent, nervous control.

Hormones are natural substances produced by glands in the body known as endocrine (hormonal) glands. These glands secrete a specific hormone which then enters the blood or lymph stream and is carried to organs or parts of the body on which it exerts an influence or control.

Thus hormones may be likened to the 'messages' carried in the nerves but they do not require trunks or fibres along which to travel. A hormone exerts a specific effect on a specific part, known as the target organ. It may help the reader at this point to take an example of a hormone and to describe its production, transport and effect on its target. Follicle-stimulating hormone (FSH) is produced by the pituitary gland, situated beneath the brain. On release from the gland this hormone is carried in the blood stream to the ovary. Here it stimulates the follicles to develop and increase in size.

A very important concept that must be explained to the reader at this point is that of receptors. Receptors are the minute points on the surface of cells to which a hormone must bind to be effective. We might liken the system to lock and keys: the key is the hormone and the lock is the receptor. The number of receptors on any given organ or target varies according to, for example, other hormone levels and the period in which the organ has been subjected to a particular stimulating hormone. Take the example of FSH, which can only act effectively on the follicles if suffi-cient receptors are present. The number of receptors may be increased by the presence of oestrogen. In practice, this explains the well known fact that follicles may be present in the ovary without increasing in size, even though the mare is in oestrus and FSH levels are rising. The follicles that do not increase in size may have insufficient receptors. This relationship is even more clearly demonstrated when we consider LH, the hormone that causes the follicle to rupture (ovulation) and shed the egg. An appar-

DAYS 0–5   OESTRUS
Tail up, winking, relaxed cervix, moist tract.
FSH and OEST dominant

DAY 5   OVULATION
Tail up, winking, relaxed cervix, moist tract.
LH dominant

DAY 7   DIOESTRUS
Ears back, kicking, closed cervix, dry tract
PROG dominant

DAY 13   DIOESTRUS
Mid cycle surge of FSH
PROG dominant

DAY 20   START OF NEW OESTRUS
FSH and PROST dominant

Fig 7 The oestrous cycle (FSH = follicle stimulating hormone; LH = luteinising hormone; PROG = progesterone; OEST = oestrogen; PROST = prostaglandin)

ently mature large-size follicle may be present in the ovary and we may augment the naturally rising levels of LH in the blood stream by injecting a dose of LH without actually causing ovulation. In these circumstances it is probable that the follicle, despite its apparent maturity, lacks sufficient LH receptors for this hormone to cause ovulation.

Fig 7 summarises the hormonal control of the oestrous cycle. If we enter the cycle at the beginning of oestrus we see that FSH levels are rising and that oestrogen is produced by the follicles as they develop under the influence of FSH. Oestrogen is the hormone responsible for the psychological behaviour of heat and changes in the genital tract such as the secretion of a moist fluid mucus and the presence of tall columnar epithelium on the uterine surface, etc. When the follicle is mature it ruptures under the influence of the pituitary hormone LH (luteinising hormone). This leads to the forming of a yellow body (corpus luteum) which secretes progesterone. Progesterone is the hormone responsible for the behaviour and changes that are associated with dioestrus and pregnancy. Both the follicles and the yellow body are temporary structures that have only a limited life, and in this respect they are unusual hormonal glands.

The life of a yellow body is cut short by the hormone prostaglandin (PGF$_2$ alpha). This hormone is produced in the uterus and probably reaches the ovary by way of the lymph stream. In cutting short the producing life of the yellow body (typically at fifteen or sixteen days after it is formed) the sex glands and organs are galvanised into a further oestrus, and the start of another cycle.

Let us now consider the way in which the cycle is controlled, how it is kept going and how it is stopped. It used to be thought that the pituitary gland dominated the hormonal orchestra by producing the hormones FSH and LH. Hormonal glands are usually sensitive to the levels of other hormones, sometimes those actually produced by themselves. Thus rising levels of the steroid hormones progesterone and oestrogen affect the rate at which FSH and LH are released. This response of hormonal glands is known as a feedback.

The theory was that rising levels of progesterone during dioestrus triggered the pituitary to produce FSH, and that in oestrus the rising levels of oestrogen caused the pituitary to release LH. Thus the cycle was thought to be conducted by the pituitary in response to its sensitivity to the levels of the two hormones produced by the ovaries. To a certain degree this view is still held, but in recent years it has had to be modified by new knowledge and understanding of the processes limiting the life of the yellow body, including the discovery of hormones produced in the brain, known as releasing factors. It seems that the pituitary is not the sole conductor of the orchestra but is itself under the control of substances produced in special cells in the brain. These are released on the receipt of nervous stimuli, mediated through the pathways of light leading from the eyes.

The influence of light on the breeding cycle of a mare has long been recognised. Increasing daylight hours, in spring and summer, promote sexual activity and cause mares to show typical oestrous cycles during these two seasons and not at other times. Very recent work suggests that the pituitary possesses an inherent primary rhythm which dictates the spontaneous rhythmical activity that we recognise as the oestrous cycle. This inherent rhythm is influenced and modified by light, acting through the central nervous system, by the hormones produced by the ovary, and by nutrition and other factors such as suckling a foal.

The Oestrous Cycle in Practice

In the Northern and Southern Hemispheres, mares in the natural state exhibit oestrous cycles in late spring and summer. Outside these seasons the cycles tend not to occur, and the mare is said to be anoestrus. There is thus a very marked tendency for mares to undergo a limited breeding season, ie oestrous cycles, and for sexual activity to cease during winter. This feature is much more prominent in the pony and other breeds in which selection for specific reasons, such as racing performance, is not intensive. From an evolutionary viewpoint, the ancestors of the present-day horse were more likely to survive if breeding occurred in the summer. Foals were consequently born in spring when the climate and fodder were more favourable to the mare in the latter stages of pregnancy, and to the foal after birth. The need for a restricted breeding season has been lessened by modern methods of husbandry. Brood mares can be stabled to protect them and their foals from inclement weather, and they can be fed hay and corn, thus simulating spring pasture conditions.

Once the constraints originally imposed by nature had been thus removed, the inherent tendency of individuals to undergo oestrous cycles during spring and summer was diluted with tendencies to cycle at varying times of the year. This process has been accentuated by commercial pressures dictating the selection of individuals for foals born 'early' in the year.

Individuals undergo oestrous cycles in (a) late spring and summer (the representatives of the natural state), (b) winter and early spring, (c) autumn and winter, and (d) periodically throughout the year. Category (b) accommodates the arbitrarily selected period of the Thoroughbred breeding season – in the Northern Hemisphere from 15 February to 15 July, and in the Southern Hemisphere from 12 August to 15 January. Mares in other categories may be a problem because they fail to 'cycle' when we most need to have them mated; for example, the mare that comes into heat regularly throughout the winter and then becomes anoestrus during the stud season. Mares that cycle outside the dates of the stud season challenge vets and veterinary science to modify and control the oestrous cycle. The extent to which we have responded to this challenge and our successes and failures are discussed later.

BEHAVIOUR

In oestrus a mare is receptive to the stallion. In the presence of the male, she straddles her hind legs, raises the tail, and urinates quantities of yellowish fluid with a characteristic odour. She is frequently seen everting

Mare showing typical signs of oestrus and straddling with hind limbs. Note the reaction of the male horse (teaser) showing Flehmen

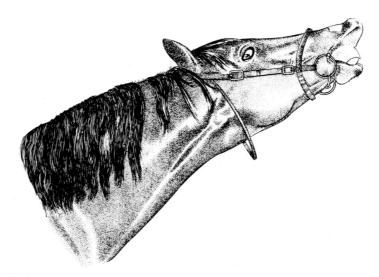

Fig 8 Flehmen's posture

the lower part of the vulva, thus exposing the clitoris. Mares on heat may stretch their head and neck and retract the upper lip in the posture known as Flehmen's (fig 8). The strength and presence of these signs vary between mares, and the determining sign of oestrus is ultimately the willingness of the mare to accept the stallion in coitus.

Dioestrous behaviour is, on the other hand, the rejection of the male by kicking, laying the ears back and biting. The vulva is shortened in length. The tail is held firmly over the perineum except when the mare urinates and it may be swished in the presence of the male.

The reader should appreciate that this description of a mare's sexual behaviour is based on our interpretation as viewed at the teasing board. Teasing is the ritual employed on the studfarm whereby mares are stimulated to display their sexual state by 'unnaturally' subjecting them to short periods of contact with the male. Under natural conditions, where the stallion runs freely with his harem, individual mares can more subtly communicate their sexual state. Visual signs may not be particularly important, although the sight of a mare 'showing' may attract the attention of her stallion from some distance. However, on approach, the stallion becomes aroused not by visual signs but by the aroma exuded by the mare in oestrus, and by the taste of her urine and secretions from the vagina and vulva. Substances which are secreted by one individual and which have an effect on the behaviour of another are known as pheromones. The stallion recognises a mare which is not in heat without the need to subject her to close attention and stimulation.

TEASING

Principles

The purpose of teasing is to cause the mare to show signs of her sexual state, ie oestrous or dioestrous behaviour. In situations where the stallion runs with his mares, his presence is a continual sexual challenge to them, and they are available to him when in oestrus. This natural situation provides a low-profile sexual relationship, free from aggressive behaviour by the male or by the female. The stallion approaches the mare when his senses of sight and smell indicate that she is in oestrus, otherwise he usually leaves her alone. There is thus no need and no place for kicking, as at the teasing board.

It is most important that we recognise teasing as an unnatural process. The greater the degree to which the mare is subjected to the male horse under the artificial circumstances of the trying board, the more likely that atypical and idiosyncratic patterns of behaviour will emerge. In particular, individuals that are in oestrus will show dioestrous behaviour, much of which may be habit-formed from repeated stimulation when the individual is in dioestrus. The artificial attention of the male causes wide variations in the mare's sexual behaviour. Horses are quick to associate and to learn from experience, good and bad. Thus if we repeatedly take them to a trying board and subject them to sexual stimulation, they may acquire the habit of showing signs of rejection, irrespective of whether or not they are in oestrus. Conversely, some individuals may become indifferent or neutral to the teaser's attentions and display no positive signs of oestrus or dioestrus.

There are other reasons why some individuals display hostility towards the teaser, such as an overriding concern for their foal or to gain freedom in the paddock. It is natural for a mare separated from her foal to have her sexual instincts subdued. This disruptive influence varies with the individual, but in general the younger the foal the more agitated the mare may become. Mares with a foal at foot for the first time may be the most disturbed. In many systems of teasing, the foal is led to the trying board and placed alongside, or in front of, the mare during the process of teasing in an attempt to reduce the mare's anxiety. Some mares may show more readily if teased after they have been set free for their daily regimen in the paddock, or as they are brought in at the end of the day. We must, of course, recognise individual variations, and the prudent studgroom takes these into account and teases the mares in several different ways each or every other day.

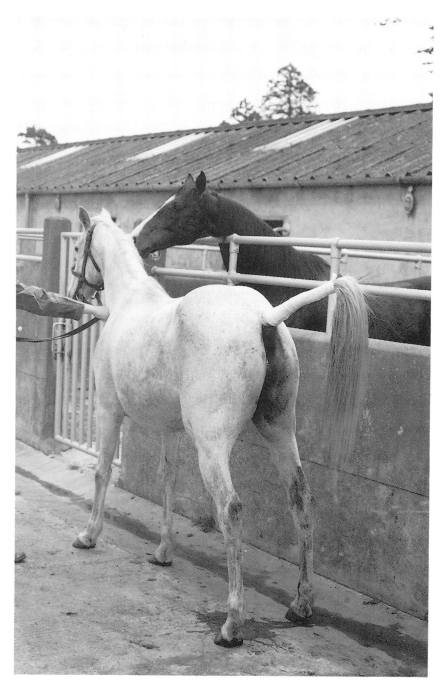

Mare indicating oestrus, and 'teaser' exhibiting Flehmen in the background

The frequency of teasing and the programme adopted are affected by the number of stud personnel available, the season of the year and the personal preference of stud managers and owners. Is there an ideal programme? The answer is *no*, for reasons already explained, because teasing is an unnatural process. We should study the individual rather than conform to dogma. In principle, mares should be teased every day during the breeding season, preferably in the afternoon or evening. Where practical, we should simulate the natural habitat as far as possible and tease in the paddocks, rather than lead mares from the stable for the purpose. This programme should be supplemented by careful observation of the mares while they are grazing, because many individuals show signs of oestrus only when relaxed in the herd. It is especially important to observe their behaviour when the male horse is in the vicinity, as when the stallion or the teaser is led past the paddock boundaries.

Of course, many studfarms do not have the resources to tease each day and many find it more convenient to stage the teasing ritual early in the morning, before the mares are turned out. Teasing at this time removes the necessity for evening teasing, when stud personnel are quite naturally taking a well-earned rest. Morning teasing also suits the daily programme of covering later in the day and of the routine veterinary examinations which may be performed each morning.

A typical studfarm routine might run as follows:

6.30am	feed stabled mares
7.00am–8.00am	teasing
8.00am–9.00am	turn out those mares stabled overnight
9.00am–10.00am	routine veterinary examinations
10.30am–11.30am	mating
11.30am–1.00pm	complete cleaning out stables; teasing mares in the paddock
2.00pm–3.00pm	start of getting mares in from paddocks, depending on time of year
3.00pm–3.30pm	second mating period of the day
3.30pm–5.00pm	complete leading in from paddocks and feeding

There are of course many variations to this type of routine, depending on the circumstances of the studfarm. A relatively short working day (nine hours at the most) has to be apportioned for a variety of endeavours, and teasing can only play a limited part of the whole. The teasing programme

Mare showing typical signs of oestrus, raising the tail, and reluctant to leave the teaser

may have to be adapted in a number of ways. For example, the proportion of mares to the number of stud personnel, the season of the year, the current climate and the proportion of foaling to barren mares all influence decisions; and during each breeding season there is a changing pattern of pressures on stud management. At times it may not be practical to tease mares every day, and an every-other-day routine may have to be adopted. Another way of reducing the pressure is to keep an accurate account of the teasing behaviour of mares. It is thereby possible to anticipate periods when teasing a particular individual is less important. For instance, a mare that has recently been mated and found at the veterinary examination to have ovulated can be expected to remain out of oestrus for at least fourteen days. Nevertheless, we have to guard against the individuals who return into oestrus shortly after having gone out of heat. These mares may be missed because they are not being teased at that particular time.

The most efficient minimum programme is to observe every mare each day and to subject her to sexual stimulation at least every other day, and preferably every day. The teasing programme must be closely integrated with the veterinary examinations which are now routine on most Thoroughbred studfarms. The success of any teasing programme depends as much on the intelligent and intuitive approach of studgrooms and their colleagues as on the stimulation by the teaser of the mares. During the breeding season stud personnel should think positively and watch at all times for signs of oestrus, such as a lengthening of the vulva or a slight change in attitude of the mare. These signs may be translated into a reliable diagnosis by presenting the mare to the teaser and then to a veterinary examination. Successful teasing programmes depend essentially on the experience and diligence of studmen.

The teaser himself may play an important subsidiary role by showing marked interest in mares that are in oestrus and a relative lack of interest in those that are in dioestrus.

The teaser should be a male horse that is not too aggressive. Savage biting, roaring or a too vigorous approach may frighten mares, whether or not they are in oestrus. On the other hand a docile, timid teaser may fail to cause a mare of similar shyness to show true signs of her sexual state. Future developments may include such devices as tape recordings of the stallion which may be played in the presence of the mares, with or without artificial scents. Dogs can be trained to detect cows in oestrus and this too might become a useful means of supplementing, or even disposing of, the teaser in the studfarm routine.

The Trying Board

The trying board should be about 8ft (2.4m) long and 5½ft (1.7m) high. It should be padded by matting or other suitable material on the side to which the mare is presented, and there should preferably be a freely turning roller about 6in (15cm) in diameter placed along the top. The boarding should be constructed of such as two thicknesses of 2in (5cm) plywood and should be situated so that the area on either side is well drained and has a firm, non-slip surface such as chalk or ridged concrete. There must be an easy approach for the mare and ample room on the teaser's side for the attendant to control the horse by allowing him to put his head over the board or to pull him back should he become too vigorous in his attention to the mare.

When constructing a teasing area, the safety of the handlers and the

The mare being presented to the teaser at the Ashley Heath Stud, Newmarket; first, she is introduced head on and then persuaded, if necessary, to stand alongside the board

precautions against injury to the mare or teaser must be given priority. The board and area should be inspected regularly for sharp edges, protruding nails or other injury-causing defects. One of the most common accidents is caused by mares kicking over the top of the barrier and injuring their hind legs on its sharp edge. Kicking the barrier itself may result in damage to the hocks or to the lower extremities if there is no soft protective covering.

The mare is led to the board and presented to the teaser head to head. The first reactions are noted, and in this position a mare that is strongly in oestrus may straddle the hind quarters and urinate. However, she might display antagonism by putting the ears back, biting or striking out with a foreleg. These signs of rejection may indicate that the mare is in dioestrus or may be pregnant and that she would not, therefore, accept the stallion.

However, a number of mares may show aggressive behaviour at first and this diminishes with prolonged contact with the male. The mare should be encouraged to stand parallel to the teasing board so that the teaser may be allowed to work along her side from the shoulder towards the hind quarters and eventually, if the mare is in oestrus, she will allow his intentions to be directed to the perineum and will show by raising the tail, squatting and 'winking'. Mares that are showing strong signs of oestrus will usually lean towards the board and may even do so if the horse is not present. On the other hand, some individuals are 'shy' and require considerable coaxing before they will allow themselves to be aligned alongside the board. The time in which the individual is in contact with the male under the circumstances of 'trying at the board' may be crucial to success in some cases.

The trying board should be placed in such a position on the stud that mares may be readily led to it in hand, thus restricting to a minimum the time spent in leading mares to and from the board. Boards which are incorporated in the fences around paddocks are useful for teasing while mares are 'turned out'. In these cases the mares may walk across to the board and 'show' while others may stay at some distance. Of these some will be in dioestrus and some will be shy individuals in oestrus who will show if brought to the board. It is advisable, therefore, to lead mares in hand to the board when teasing from the paddock boundary.

In areas where mares are herded, use is frequently made of a crush through which they are driven and brought into contact with the teaser. This method is a variation of leading mares to the trying board and it effectively constrains the mare so that the teaser may be brought into close proximity.

However, one of the disadvantages of teasing at the trying board is that the mare is exposed to the attention of the male for only a short period of time and under conditions that are quite different from the natural circumstances of the stallion running loose with his harem and thus being constantly present.

The Teaser in Constant Contact with the Mares

An alternative to the method of planned teasing as described above is to have the mares in constant contact with the male. There are a number of different ways that this can be achieved. Placing a pony stallion in a small paddock or yard which abuts on two to four larger paddocks in which the mares are kept is a method especially favoured in the Southern

Hemisphere, in Australia and New Zealand. The perimeter fencing of the male horse's compound must, of course, be constructed with care to allow adequate contact yet no chance of escape. If the male horse is small relative to the mare, for example a Shetland male and Thoroughbred mare, the risk incurred of the horse's escaping and mating one of the mares is correspondingly reduced. The temperament of the male horse may also influence the success of this method because some individuals settle quite readily to the constraint and separation from the mares. Because the situation is more natural, the mares become accustomed to showing at the intervening fence for periods which vary according to the strength of oestrus experienced by the particular individual. Some mares will come occasionally to the fence and show, while others keep near to the male horse throughout most of the day. The success of the method depends, as in all other methods of teasing, on the ability of attendants to observe and interpret the signs according to the circumstances and the variations occurring between individuals. The constant presence of the male in the vicinity of the herd, being more natural, is less dramatic than the enforced courtship of teasing at the trying board or other methods. A further extension of the method is to run a vasectomised male or one which is too small to mate the mares under test, and to leave these horses within the herd. The disadvantage of the vasectomised male is that this may lead to a transfer of venereal infection and it also exposes the mares to coital challenge unnecessarily (see page 132).

Teasing in the Paddock

We have already discussed the method of leading the teaser to the trying board at the boundary of the paddock. A variation of this method is to lead the teaser around the outside of the paddock, observing the behaviour of the mares and their reaction to his presence. It is customary in many studs in the UK to exercise the stallion in this way and for the stallion man to report on the behaviour of mares as he walks the stallion past the paddocks in which they graze. Sometimes the horse may be taken to the fence to sniff or nudge the mare on the other side. However, this manoeuvre is not recommended, because of the danger of the mare or the stallion striking into the fence thereby sustaining injury.

The behaviour of horses is based on herd instincts, and social bonds play an important role in determining their behavioural patterns. Isolation is unnatural, and contact is a reciprocal and interacting stimulant. Thus the presence of other members of the herd in the paddock may provoke sexual displays which would not occur in the isolated environment of the loose box. In the paddock one mare may be seen showing oestrus signs to another, and sometimes mares may pair, even to the extent that their sexual cycles tend to run concurrently. It is rewarding for studmen to observe carefully the behaviour of mares at pasture when they are in contact with each other and, in the context of the stallion being exercised past the paddock, in the presence of the male. Experienced studmen may deduce that the mare is in oestrus by her attitude in the paddock. For example, she may display the Flehmen posture, straddle without passing urine or indulge in mutual grooming at the same time as raising the tail and, perhaps, everting the clitoris. Changes in behaviour such as restlessness or whinnying may cause the observant studman to suspect that an individual may be in oestrus. Much of this 'free' teasing rests on the intuitive sense of

Two mares showing interest in the teaser who is being brought to the board incorporated in the paddock fencing

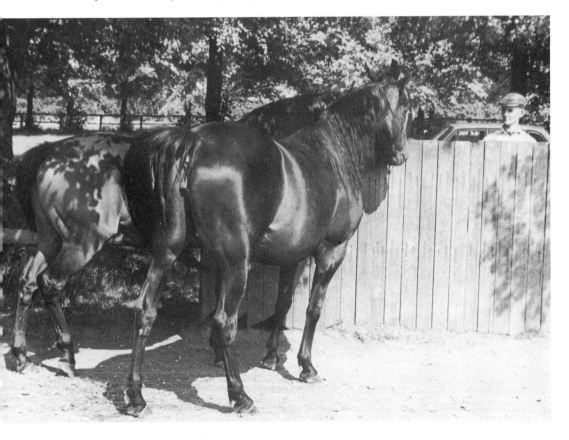

the onlooker; a written description lacks the authenticity that can be brought by experience and an intelligent application of knowledge of the sexual and physiological functions underlying the oestrous cycle.

Other Methods of Teasing

There are really no limits to the different means by which the male and female horse are brought into contact for the purpose of promoting sexual signs. The teaser may be brought to the loose box door or walked through the paddock. He may be ridden through the stud or led in hand. However, whatever method is chosen, the safety of the handlers and of the male and female horses must receive first priority. The methods touched on here have obvious risks because much will depend on the nature of the stallion and, to a lesser degree, on that of the mares. To ride many teasers might be foolish, and to take a male horse into a paddock of mares could involve serious injury to man or beast, although this method has been used successfully when a particularly quiet teaser is employed.

INFLUENCES AND VARIATIONS IN THE CYCLE

Influences

Sexual activity is influenced by season, nutrition and climate. Light, warmth, and feed containing good-quality protein combine to stimulate sexual activity. These are, of course, the conditions of spring and summer. Decreasing daylight hours and reduced warmth and dietary intake combine to have the opposite effect, to suppress the oestrous cycle. Sexual inactivity may occur at any period of the year but is more common in autumn, winter and early spring.

The spring and summer are scientifically known as the ovulatory and the autumn and winter as the anovulatory seasons. This refers to the fact that ova (eggs) are produced during oestrus. When mares switch from the anovulatory to the ovulatory season, usually in spring, and from the ovulatory to the anovulatory season, usually in autumn, the oestrous cycle may depart from the five-day oestrus plus fifteen-day dioestrus referred to earlier as typical. In the changeover there are many individuals that have prolonged oestrus ranging from ten to twenty days, and sometimes even longer. In some cases these extended oestrous periods end with the development of a follicle and ovulation, but in many instances the mare goes out of oestrus without actually ovulating.

Anoestrus sometimes occurs in mares that are suckling a foal because, it is thought, the pituitary is producing the milk hormone prolactin which suppresses the release of pituitary hormones FSH and LH; thus lactational anoestrus may be difficult to overcome without weaning the foal.

Variations

We have already considered some of the ways in which the oestrous cycle varies according to the individual and the season. It is of the utmost importance in practice that we understand these peculiarities as they affect both the interpretation of sexual signs at the trying board and the understanding of these signs in relation to what is, or is not, occurring in the ovaries and in the physiological make-up of the mare.

On page 71 we discussed the phenomenon of receptors through which hormones exert their action, as in the manner of a key turning a lock. If there are insufficient receptors, the action of the hormone, however high its level in the bloodstream, is correspondingly diminished. Variations in the oestrous cycle may result from insufficient receptors being present. For example, a mare that does not allow a stallion to mate her and appears to be out of heat when all other signs (zero blood progesterone levels, follicles present in the ovary) indicate otherwise, may be suffering from insufficient receptors in any of the target organs (eg uterus and genital tract) where oestrogen exerts its action. This also may apply to the brain where the psychological effects of oestrogen promote a feeling of receptivity.

1 *Oestrus without ovulation (spring oestrus)* The anovulatory oestrus may be encountered at any time of year, but is more usual during the long oestrous periods experienced in winter and early spring, as the mare changes from anoestrus to cycling activity. The ovaries may remain inactive or develop follicles, none of which matures sufficiently to ovulate. There is no means of distinguishing an anovulatory oestrus from one in which ovulation occurs without performing a rectal examination or measuring progesterone levels in the blood.

2 *Post-oestrus ovulations* Some mares may go 'out of oestrus' one or two days before they ovulate. This means that the service of the stallion may bring sperm into the mare too soon before ovulation for them to survive (see optimal time of mating, page 91).

3 *Luteinisation of the follicle without ovulation* It seems that a 'ripe' follicle may sometimes become luteinised, ie produce progesterone without actually rupturing.

4 *Mid-cycle ovulations* Ovulations occurring during dioestrus are quite common. Their significance in practice is difficult to assess in the light of present knowledge. The LH peak in the mare occurs after, rather than coincidentally with, the time of ovulation and may therefore cause more than one follicle to ovulate. The second ovulation may result in twins being conceived. Ovulations in mid-cycle may interfere with prostaglandin treatment. If the hormone is administered within four or five days of ovulation it will not 'kill' the yellow body which is resistant to its action for this period. Secondary ovulations do not appear to prolong the dioestrus in which they occur. The natural release of prostaglandin from the uterus on the fifteenth or sixteenth day of dioestrus kills both yellow bodies. However, if the second ovulation occurred within four days of the natural release we would then expect to have a long dioestrous period.

5 *Prolonged dioestrus* The functional life of the yellow body may be prolonged beyond the usual fifteen to twenty days. The individual fails to 'come back' into oestrus, and it may be assumed erroneously to be pregnant. These yellow bodies may last for double or treble the usual period, or even continue their activity for months. The reason for their exceptional life span is unknown. In some cases it may be due to the presence of a fertilised egg in the uterus, which does not live for more than a few days. Mares become programmed for pregnancy at about fourteen days from conception, but it is not known what other mechanisms determine that the yellow body formed at the time of ovulation is switched into the yellow body of pregnancy, thus continuing its life beyond the usual life span. If, however, pregnancy has proceeded beyond this point it may be one explanation for a prolonged dioestrus. Some mares experience prolonged dioestrus when they have not been mated, and clearly the early death of the conceptus is by no means the only reason for this phenomenon.

6 *Silent heat* There are two forms of this condition. In one, the mare undergoes physiological oestrus, ie she develops the typical changes in the ovary and genital tract, including ovulation, without exhibiting behavioural signs. However, this is usually a matter of misinterpretation on the part of those responsible for teasing the mares of the signs

displayed by the individual at the trying board. A substantial number of mares exhibit contradictory behaviour in the circumstances of artificial sexual stimulation, and it is hardly surprising that many individuals may fool us as to their true sexual state. We are able to distinguish these mares by observing the cervix through a speculum or by palpating the uterus and cervix through the rectum. If discovered in time, all of these mares, with very few exceptions, would take the stallion if presented to him. Indeed, many display outward signs of oestrus when the stallion approaches.

The second type of silent heat involves the physiological and behavioural components of the cycle being out of step with each other. The mare may ovulate before showing signs of heat or without the usual changes occurring in the tract. In both instances the mare is non-receptive to the stallion during the important period for coitus, ie twenty-four hours prior to ovulation. The cervix at this time is closed and pale and the mare would kick the stallion if presented to him. This situation is similar to the mid-cycle ovulations already discussed, but differs in that they occur at that time of the cycle when an oestrous period is due. It has been shown experimentally that some of these mares can conceive if they are insemi- nated artificially at the time of ovulation.

7 *Oestrous behaviour in mares not in physiological oestrus* This condition could be regarded as the opposite to silent heat. It may occur when the mare is pregnant, especially if she is pregnant for the first time. It may also be found in non-pregnant mares during the late winter and early spring. The mare displays signs suggesting that she would accept the stallion, but in the event she kicks and rejects him on approach. There are, however, a small number of mares who may even accept the stallion in these circumstances. If we examine the cervix we find it is closed and pale in the typical dioestrus or pregnant condition. Ovulation may or may not occur but there is usually some follicular activity in the ovaries. If a mare is pregnant and mated there may be one of three outcomes: abortion; the pregnancy is unaffected; or a second foal is conceived and twins develop. In the author's experience, the most common result is for pregnancy to continue. However, the last service date of that particular pregnancy is erroneously recorded and the mare foals three or six weeks prior to the expected time. Abortion has been recorded following service of a pregnant mare but the conception of twins is the rarest result of the three.

8 *Aggressive behaviour of the mare when in oestrus* Some individuals may refuse to accept the stallion even though they appear to be in oestrus

from their signs at the trying board and from a visual inspection of the cervix. They are usually maiden mares or those who have a foal at foot for the first time. This is not a common condition and is presumably the result of environmental conditions of breeding in hand, although some individuals appear to inherit this type of behaviour.

KNOWLEDGE APPLIED IN PRACTICE

In this chapter, I have reviewed the knowledge on which our understanding of oestrous function and behaviour is based. Such a knowledge and such an understanding must be the cornerstone of our approach to the breeding of horses. In practice all of us – vets and studmen – are required to make judgements and to take decisions consistent with the facts as we know them. One of the problems is that our knowledge and understanding is inevitably limited in the sense that there is always more to be discovered about any given subject. Further, the knowledge we do acquire from year to year cannot always be understood nor put to contemporary use. For example, we have already seen that the life of the yellow body may be prolonged under certain circumstances. If we knew how to turn this particular switch we could do so to keep fillies in training out of oestrus or to keep mares from cycling at times of the year when we wish them to be sexually quiescent, thereby perhaps enhancing their activity when we most require it.

We must not slip into the error of assuming that human assistance is essential for the breeding of horses. Place a male horse among a harem of females, give them ample feed and reasonable protection from extremes of climate, and conception rates will be reasonably high. If mares that breed less than every other year or, even, less than every year, are culled from the harem the percentage rate will increase, provided of course the male is fertile. The same result may be found in harems where breeding in hand is practised, although in this case the acuity of observation and diligence of studgrooms, etc, plays an essential part in achieving respectable rates of conception.

Artificial aids and a deep understanding of the sexual functions are only required when male and female breeding stock are selected for such endpoints as racing performance to the exclusion of breeding efficiency. Perhaps the most important lesson that we can learn from our present knowledge of the oestrous cycle is that patience is an essential ingredient for success. One of the most common myths in our attitude towards the oestrous cycle is to describe aberrations as *abnormal*. It is absurd to designate a natural physiological variation as if it were some pathological process or abnormal condition. The physiological function of the oestrous cycle obviously encompasses a wide range of variations and if these are to be expected they can be anticipated, thus providing the spectator with a heightened awareness of the behavioural nuances that may be encountered. This in turn should lead to better results in practice and to less emphasis on therapy for naturally occurring states. There are important lessons to be learnt from our present state of knowledge and these are summarised as follows.

The seasons of the year may influence the sexual activity of some individuals more than others. Owners may save time, money and frustration by using veterinary expertise to identify the particular state of any given mare in their possession according to the month of the year and against the background of their requirements. For example, if an owner possesses a barren mare, he can send her away to the studfarm for covering on 15 February, the start of the official breeding season. But if he is wise he will have her examined by the vet to ascertain if she is in deep winter anoestrus. He can then retain her on his own farm at less cost until such time as she starts 'cycling' and is ready for service. By these means an owner can keep charges down to the cost of a relatively small fee for a veterinary diagnosis. There are many other ways in which veterinarians can help to reduce the overheads of mare owners, but there are limits which must be fully appreciated.

The following are some of the ways in which we can put our knowledge into practice to assist rather than to counteract the natural events.

Choosing the Time of Mating

There is an optimal time for coitus which ensures a minimal number of ripened sperm in the oviduct prior to, rather than after, ovulation has occurred. The basis for choosing the occasion for mating rests on our knowledge that the mare's reproductive tract is a hostile environment for sperm, the hostility varying according to the presence of infection and metabolic changes affecting acidity and the nature of the mucous lining. Immunological and hormonal changes may affect the survival and mobility of the sperm, reducing the numbers available for fertilisation. The sperm must reach the oviduct and they need a number of hours to ripen before they are capable of fertilising the egg. It seems that, although only one sperm fertilises the egg, it is necessary to have a large number present in the Fallopian tube, perhaps to assist by producing enzymes. The minimal number for this process is not known but it could involve hundreds or thousands. Once the egg is shed at ovulation it becomes increasingly resistant to being fertilised and after twenty-four hours it is virtually 'unfertilisable'.

The stallion's influence includes the number of live normal sperm contained in the ejaculate deposited in the uterus at coitus. A horse with a relatively low number of normal live sperm (see page 127) will provide a lower number of sperm and, because in all cases live sperm numbers decrease hourly as the 'weaker' ones die or lose their mobility, the number of hours prior to ovulation that coitus occurs is all-important. The optimal period for coitus is estimated at about six to twenty-four hours prior to ovulation. It is customary to allow forty-eight hours to elapse before a second service is given, but this interval is based on the tradi-tional view that the life span of the sperm ejaculated by the average Thoroughbred stallion lasts for two days. In practice, there is a wide variation, some stallions being potent for barely more than twenty-four hours and some up to seven days. This potency will, of course, be affected by the state of the mare's genital tract.

The Behaviour of the Mare Relative to the Time of Ovulation

Ovulation is not predictable in terms of the start of oestrus but only in retrospect from the end of oestrus. In the absence of veterinary examina-tions in which the ovaries are palpated, there is no exact means of determining the optimal occasion for coitus, ie six to twenty-four hours prior to ovulation. However, experienced studgrooms can, in certain indi-viduals, discern a change in sexual behaviour close to their point of ovulation. This usually relates to the strength of signs displayed when the mare is teased; the term 'very well in' aptly describes the situation in these cases. However, this does not apply to those mares who show very strong signs throughout the whole of the oestrus; in these there is no possibility of judging the point of ovulation. In fact some of these individuals are so much affected by oestrus that the signs continue for some days after they have ovulated. It is not, therefore, practical to base one's covering programme entirely on the signs shown by the mare. However, some mares may show a similar pattern from year to year, and the history of the mare's sexual behaviour may sometimes be helpful in judging the optimal time of mating in the absence of veterinary assistance. Hence the value of keeping charts containing facts about the length of oestrus and covering dates in each stud season.

Sexual Quiescence

The absence of oestrous activity may indicate that the mare is pregnant, in a state of prolonged dioestrus or anoestrus.

Irregular Cycles

Short dioestrous periods (under twelve days) may indicate that the mare did not ovulate in the previous oestrus period or that there are pathological reasons which are shortening the life span of the yellow body, such as uterine infection.

RATIONALE OF VETERINARY EXAMINATIONS

The veterinary examination is an adjunct to the teasing programme. By palpating the ovaries and the uterus per rectum and/or by examining the cervix through a speculum the vet is able to determine the sexual state of the mare at any given time. In this way the signs of sexual behaviour may be confirmed and a diagnosis made of the optimal time of mating or, at the other end of the scale, whether or not the individual is pregnant.

There has been criticism of the over-use of the veterinary examination

Rectal and vaginal examinations help to confirm the sexual status of the mare at any given time

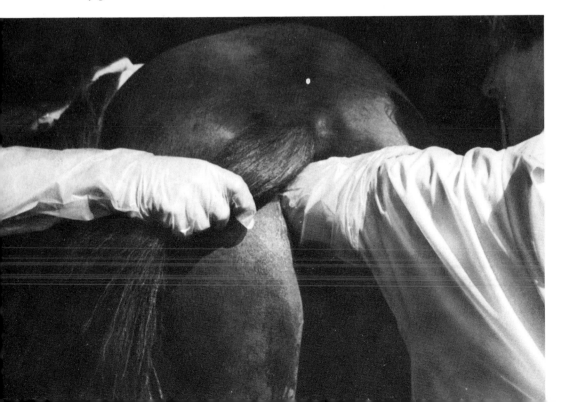

on studfarms. Some owners complain that having sent, say, a healthy maiden mare to a studfarm, these individuals are examined many times, and veterinary fees charged which in aggregate may amount to one or two weeks' keep charges. Why, these owners ask, did my mare require so much attention? To answer this and other similar questions it is necessary to appreciate the co-operative nature of the veterinary examinations in the breeding programme. In the first place, we must distinguish between managerial and veterinary aims.

Managerial aims

(1) To determine the sexual state of oestrus, dioestrus or early pregnancy (nineteen to forty days).
(2) To determine the optimal time for mating relative to the needs for mating other mares where more than two are involved on any one day in the stud season.
(3) To estimate the optimal time of coitus so that this occurs close to the time of ovulation.

Veterinary aims

(1) To diagnose the presence of infection and treat accordingly.
(2) To reduce the number of services in mares susceptible to infection (see page 132).
(3) To diagnose functional and pathological variations which might be responsible for a mare failing to conceive.
(4) To confirm whether or not pregnancy has occurred at the earliest opportunity and to diagnose and eliminate twins (see page 155).

It is not always possible to distinguish between the managerial and veterinary motives for the examination in any particular case. For example, a mare that is showing indeterminate signs at the teasing board (leaning towards the board but squealing and kicking at the same time) requires a veterinary examination to confirm that she is, in fact, in oestrus. This is essential in the interests of the stallion owner(s) to avoid the risk of injury. The examination might reveal that a follicle is present in the ovary and likely to ovulate within forty-eight hours. Management, therefore, has exact information and can plan accordingly. However, at the time of examination, the vet finds evidence that the mare is susceptible to infection and requires some preventive treatment to increase the chances of a successful mating. This diagnosis and the subsequent therapy is a veterinary function related to the individual herself and not to the management of the stud in general nor the interests of the stallion owner(s).

In this example we see the interdependence of managerial and veterinary responsibilities. But let us take the case to a further extreme. Supposing that at the routine examination this particular mare was found to be infected with the organism CEM or of *Klebsiella*. Both these microbes are capable of venereal spread, and the infected mare thus poses a risk to the stallion and through the stallion to other mares should he be mated. The responsibility for avoiding this particular hazard rests on a veterinary diagnosis followed by a managerial decision, ie the mare should not be mated until she has been cleared of the infection. There are numerous variations on these themes, many with less clear-cut distinctions. For example, a mare that has a prolonged oestrus, lasting two or three weeks, may be examined repeatedly on behalf of management so as to avoid unnecessary matings. When a follicle develops in the ovary and the mare is thus ready for mating, the diagnosis — that coitus at that time will provide a greater likelihood of conception occurring — is one made in the interests of the mare owner. From a veterinarian's viewpoint, the knowledge of impending ovulation may be essential to helping the mare to conceive. It is thus part of the treatment.

It is perhaps in the costing of the examinations that most dissension about their necessity arises. The question that is rarely answered is to what degree the responsibility should be that of management of the stallion and to what extent the mare owner. The bloodstock industry should come to some decision. All that can be achieved here is to lay down the guidelines on which such a decision might be made by individual studfarms or as a general policy acceptable to the industry as a whole.

Handling the Mare

Vets rely on the handlers of mares to facilitate the examination. The aim is speed with efficiency and safety. The quicker the examination, consistent with accuracy, the less the time that personnel will be occupied by this routine. Where few numbers are involved this may not be important but, on a large studfarm, where twenty or thirty mares are examined in one session, a saving of two or three minutes per mare can save considerable man-hours per day, especially because at least two attendants are required for handling and a third for recording results.

There are two basic methods of handling mares for the veterinary

Restraint by means of a twitch is the traditional approach to control of the mare during mating

gynaecological examination, namely stocks and around a door post (fig 9). Stocks are favoured in the Southern Hemisphere, many European countries and in North America. Or mares may be examined in their stalls and this was traditionally the custom in many Thoroughbred breeding centres, such as Newmarket. However, nowadays an increasing number of studfarms have stocks. The disadvantage of stocks is that mares have to be led in hand and this is time-consuming, although providing the benefit of a central point where equipment may be stored and undercover facilities provided. To move from box to box is, perhaps, a case of the mountain coming to Mohammed, but the system works reasonably well. The method stems in particular from the fact that the design of studfarms tended to disperse boxes towards the periphery of the premises, in an effort to place them near to paddocks from which mares could be taken without having to lead them too long a distance. It is necessary in these situations to arrange stocks in each range of boxes, and has therefore been found more convenient to use the round-the-door method. There are, from the vet's viewpoint, certain advantages and disadvantages in each method. On balance, the examination made in stocks, under cover, provides

Fig 9 The veterinary examination may be made round a door post for protection: note the way the attendants are positioning and restraining the mare for the vet. If the mare has a foal at foot it is best held in front of her or allowed loose and trapped between the mare and the corner of the stall on the right-hand side of the mare

maximum cleanliness, lighting and freedom from risk of injury. The examination in a box is a less orderly affair and one in which all concerned are more at the mercy of the mare's behaviour and, if the box opens to the outside, of the prevailing weather.

The advantage of a system for examining mares as one finds them, so to speak, is that the technique may be used in other circumstances, such as medical diagnosis of disease, and in circumstances where stocks are not available.

It is essential to have the mare restrained by two attendants, one holding the head and one the tail. The hind quarters of the mare are manoeuvred to a position so that they are at the opening of the door but, should the mare kick, the hind legs pass across the doorway rather than directly towards the outside. Most mares may be restrained by means of a halter or head collar and rope. However, some individuals require the use of a bridle and snaffle bit or chifney. Those responsible should make sure to disinfect the bit before using it on another mare. The habit of 'bridling' many mares for the routine veterinary examination is to be avoided because it is an obvious means of transferring infections, such as strangles and viral diseases, from one mare to another.

A skin or muzzle twitch hold may be a helpful and necessary means of restraint for fractious individuals. The raising of one foreleg is sometimes practised, although there is some risk that the mare will plunge and fall; a twitch hold is preferable. It is important that the handlers work in close harmony with the vet, adapting to any system or modification of method that the individual may prefer. The vet should be warned in advance of any known vicious or fractious animal before starting an examination.

Nowadays the examinations are complicated by requirements for swabbing and carrying out other techniques; and an extra pair of hands is essential. There is also the need for strict cleanliness and hygiene. The environment of the examination should be free from dust, contain adequate lighting, be protected from draughts and have ready access to hot and cold water. During the CEM outbreak of 1977 running water was advocated for washing the mare's perineum, and various devices developed for the purpose. The water employed should contain a mild disinfectant. Care should be exercised that neither the water be too hot nor the disinfectant too concentrated as this may scald the vulva. Depending on the size of the studfarm, the vet may prefer to have protective clothing including a smock and boots kept on the premises.

The instruments for examination should be kept on the premises and a sterile speculum used for the examination of each mare. The results must be recorded at the time of examination. There are many different methods favoured by vets according to their own experience. Symbols of lettering and/or numbers are in most common usage, but drawings and even written descriptions are sometimes employed. The important need is for the meaning and interpretation of the symbols to be clearly understood and comparable from one examination to another. By this means we can examine the history of a mare's sexual activity and genital status in retrospect and have available to us a history on which to base current diagnosis. The record should also allow other people to interpret the findings, when those who made them are not available. In theory it would be ideal to have a standardised method of recording used by everyone in the bloodstock industry, but this ideal is never likely to be achieved. In any case, the actual findings are largely subjective and vary from one operator to the next. For example, the diameter of a follicle has to be estimated by feel, not a very accurate method, and findings may differ by as much as 0.4in (1cm) between two experienced vets. In practice most vets denominate a figure to serve as an index rather than a measurement. This is acceptable provided the management can rely on consistency of the individual's results or on the advice consequent on the findings. It is generally accepted that a follicle of $1\frac{1}{5}$in (3cm) diameter or more is ready to ovulate, although those of less diameter often do so. Size alone cannot, therefore, be taken as a guide. Consistency, tension and position in the ovary are also important factors and should be recorded. One system of recording ovarian activity is to designate the poles of the left ovary as a and b, and those of the right as c and d. In this way, the position of the follicle may be recorded together with the size.

The cervical examination may be conveniently recorded by a numerical system of, say, 1 to 6, with variations of consistency, colour and moisture: category 1, for example, denoting a wide open oestrus-type cervix and category 6 a tightly closed, dry dioestrus or pregnant state; numbers 2 and 3 being signs of oestrus and 4 and 5 those of dioestrus. Abbreviations of full descriptions may be used for other physical findings, such as urine pooling on the floor of the vagina, the presence of pus, etc.

The introduction of ultrasound scanning in recent years has made it possible to provide a very accurate description of the contents of an ovary in terms of size, shape and position of the follicles present. Further, the condition of the uterus can be assessed and fluid, oedema or cysts identified. This subject is discussed in more detail in chapter 14.

Artificial Means of Controlling the Oestrous Cycle

Controlling the oestrous cycle implies, in effect, causing mares to come into heat and to ovulate when we want them to do so. There are managerial and veterinary reasons for this which will be discussed separately, although the means to the end are similar, if not identical. First we must define the objects of control. They are:

1 To bring mares into oestrus and to cause them to ovulate in the months we require to mate them.
2 To produce a sufficient number of ovulations during the breeding season to give the maximum chance of conception, given existing conditions of breeding in hand and the vagaries of modern stud management. If conception occurs on the first heat period in which the mare is mated there is obviously no need for further ovulations. But if, as often is the case, a mare does not conceive, the sooner she is returned to oestrus to provide another chance the better.
3 Breeding in hand requires mating on the predetermined occasion. It may thus be desirable to hasten (or delay) ovulation so that coitus occurs at the optimal time, ie just prior to the shedding of the egg.
4 Controlling the number of ovulations within an oestrous period would be a very worthwhile method of avoiding twins. However, at the present time, there is no appropriate means of achieving this end (see Twins, page 271).

The conditions we have to overcome are:

(a) True anoestrus, in which the ovaries are inactive, blood progesterone levels are zero and pituitary FSH and LH levels minimal.
(b) Prolonged dioestrus, in which the yellow body continues to function after the normal fifteen-day period, thereby preventing the mare from developing oestrus. In these cases there is progesterone present in the blood.
(c) Returning normally cycling individuals to oestrus should they not be mated (ie missed) for various reasons.
(d) Anovulatory oestrus, in which the mare shows signs of oestrus but does not ovulate, even though follicles may develop in the ovary. These 'failures' may be due to a deficiency of LH and/or the fact that the follicle has not matured due to some previous happening. Let me explain this by an example. Recent work has shown that an FSH surge occurs mid-cycle and it has been suggested that this may be needed to prime the follicle for

the subsequent heat period. If this priming does not occur the follicle might not develop to maturity and might therefore fail to ovulate, even though the pituitary had produced sufficient LH for the purpose.

(e) A semi-luteinised follicle, ie one that has partially but not completely ovulated, may form in the ovary and block subsequent oestrus.

We have only limited means at our disposal for controlling the oestrous cycle, and consequently our success is correspondingly incomplete. Further, our understanding of the subject is based largely on fragmentary information coming from clinical observations and isolated research programmes. But the measurement of hormone levels in the mare's blood provides only a restricted view of what is happening at the producing target organs.

Let us turn to the artificial aids that we have at our disposal at the present time and make some comment on their efficacy in practice.

(a) GNRH (gonadatrophin-releasing hormone). This, as its name implies, is the hormone produced by the brain cells that releases the gonad-stimulating hormones of the pituitary. It has an FSH and an LH component. There are synthetic compounds of this hormone available commercially and they may be used in an attempt to promote the growth of follicles in a mare showing oestrus without ovarian activity; or in anoestrous mares to encourage them to start cycling. GNRH is best administered in the form that mimics natural output, ie in a pulsatar fashion. If we administer GNRH in a single bolus dose it has very little effect, but if given in the form of an implant so that there is a slow release over several days the effect may be quite successful; follicles are stimulated to grow and mature within the ovaries, and eventually one of these will ovulate.

(b) Artificial lighting to extend the daylight hours is the most effective way of causing anoestrous mares to cycle early in the year. There are numerous different ways of applying this technique. In general, a 200-watt bulb placed 12ft (3.6m) or so above the mare in a loose box and left on for six hours after nightfall is sufficient to arouse mares from inactivity, if the programme is extended over about two months. Increasing warmth and protein levels in food are useful adjuncts to artificial lighting programmes.

(c) Progesterone compounds injected, or related compounds placed in the feed daily over a period of ten days or so, can be used to stimulate some mares from anoestrus into oestrus, and those in prolonged oestrus

towards follicular development and ovulation. The increase of progesterone in the mare's blood stream affects the pituitary by inhibiting the
release of FSH and LH. While the progesterone is circulating the pituitary
stops releasing FSH and LH. When the course of progesterone is stopped
this pent-up material is released and effectively promotes a true oestrous
state. This explanation assumes that small amounts of FSH and LH would
normally be 'leaked' from the pituitary over the ten-day period of the
course and be insufficient for any worthwhile effect on the ovaries. An
alternative programme, favoured in North America, is to inject a combination of progesterone and oestrogen daily for ten days.

(d) Luteinising hormone (LH) has been used for many years to cause
ovulation in mares that have a ripe follicle in their ovary. This enables us
to cut short oestrus and arrange coitus so that it occurs within forty-eight
hours or less prior to ovulation. The limitation of the therapy is that the
follicle must be one destined to ovulate. We have a ready source of a
hormone with LH activity in pregnant women's urine. This LH comes
from the human placenta or chorion and hence is called human chorionic
gonadatrophin (HCG).

There has been some concern as to whether the repeated use of HCG
might act in a similar way to a vaccine and cause antibodies to form,
thereby nullifying the effect of the HCG and actually diminishing the
effectiveness of the mare's own pituitary LH. Recent work suggests that
antibodies are formed but do not interfere with the mare's own LH nor do
they prevent HCG from being effective when injected in the right circumstances. The use of HCG, by facilitating ovulation, could increase the risk
of twin ovulations and of twin conceptions, but there is good evidence that
twin ovulations result in twin conceptions only in a very small percentage
of cases. In a survey carried out by the author over a number of years the
incidence of twins in mares given HCG was no more than in those which
were not. GNRH may also be used to cause ovulation because it has an
LH-releasing component. There is some subjective evidence that it may
cause the ovulation of smaller and rather less well-developed follicles
than is the case with HCG. A study performed in Newmarket showed that
the injection of HCG compared with the administration of GNRH or a
saline control resulted in ovulations occurring between about thirty-six
and forty-eight hours post injection. It seems that injecting HCG delayed
ovulation for about twenty-four hours but hastened its occurrence in a
twenty-four to forty-eight-hour period compared with the controls in
GNRH where ovulations occurred over a much longer period.

4

THE SEXUAL FUNCTIONS
OF THE STALLION

It is curious how often one hears vets, and others intimately concerned with breeding horses, make such remarks as 'we got that mare in foal with difficulty', or 'such and such a mare produced a colt or filly foal'. It is comparatively rare to hear the stallion given much credit for his part in the affair of 'getting' his mare in foal or of being responsible for the product of foaling. Thus, it appears, we all take credit for the finished article before admitting the essential role of the stallion in the reproductive process. Yet the stallion subscribes fifty per cent of the genetic material of the individual at conception; and his influence on fertility is as important as that of the mare he serves. When we discuss the subject of infertility (see page 243) we shall find that there is a close interrelationship between the fertility of the mare and the stallion, a subfertile horse limiting the capacity of a subfertile mare to conceive.

ANATOMY

The sex organs of the stallion comprise the two testes and the equivalent of the mare's genital tract, ie the tube that provides access between the gonads and the outside world. However, in the case of the stallion, there are ducts leading from the gonads to the urethra which form a common exit in the penis for the discharge of urine and the products of the testes (fig 10). The accessory glands contribute special substances and fluids to the spermatozoa as they travel through the ducts thus forming the milky white secretion called semen.

The Testes, Epididymis and Spermatic Cord

The testes are roughly oval-shaped and compressed from side to side thus presenting two surfaces, two borders and two extremities. The testes, in the mature horse, lie in the scrotum with the lower border free and the upper border attached to a membrane containing the epididymis.

Each testis is about 5in (12cm) long, 3in (7cm) from the upper to the

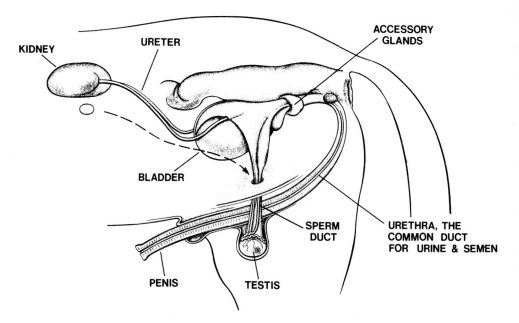

Fig 10 The stallion's genital organs viewed from the left side. The arrow repre-
sents the path down which the testis migrates during the latter part of foetal
development, passing through the inguinal ring and down into the scrotum

lower border, 2in (5cm) wide, and weighs about 10oz (300g). The testes
vary, however, with the individual and its maturity, and the left testis is
often somewhat larger than the right.

The epididymis adheres to the upper border of the testis and somewhat
overlaps its lateral surface. Its front end is enlarged and called the head, its
hind end is enlarged, but to a lesser extent, and is called the tail. The inter-
mediate middle part is known as the body. The head is closely connected
to the ducts that emerge from the testes and the tail is continued by the
major duct known as the *ductus deferens* which runs in the spermatic cord
and carries the sperm upwards on their way from the testes to the urethra.

The spermatic cord extends from the inguinal ring, in the abdominal
muscle above, to the testis in the scrotum, below. It contains an artery,
veins, lymph vessels, nerves, the *ductus deferens,* a muscle and an outer
fine membrane which extends to cover the epididymis and testis.

Beneath this membrane, as it passes round the testis, is a strong capsule
which, when cut, exposes the soft and reddish-grey coloured substance of
the gland beneath. The interior of the testis is divided into compartments
by septa of connective tissue and striped muscle. The compartments or

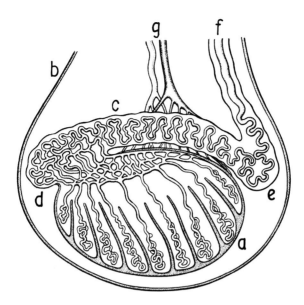

Fig 11 The testis (a) in the scrotum (b) indicating the blind-ended tubes that join to form the epididymis (c) with a 'head' (d) and 'tail' (e). Sperm pass from the epididymis into the vas deferens (f) that, together with the blood vessels (g), forms the spermatic cord

lobules consist of the seminiferous tubules, which are very small winding ducts that end blindly at one end and, at the other, unite with other tubules to form larger straight tubules (fig 11). In summary, sperm are produced in the testis, pass into the epididymis for storage and 'ripening' and then, at the time of ejaculation, are propelled through the *vas deferens* and urethra with the addition of seminal fluid and gel produced by the accessory glands.

The Scrotum

The scrotum, which houses each testis, consists on the outside of thin and elastic skin containing sebaceous and sweat glands. The middle lining is formed by elastic tissue and unstriped muscle. Beneath this there is a layer of connective tissue lined on the inside by a layer of the *tunica vaginalis,* the fine membrane which has its counterpart lining the testis, thus enabling this organ to slide freely within the scrotum. The testes can be withdrawn, partly by contraction of the muscle in the spermatic cord and partly by that in the scrotum. Thus on cold days, during exercise or at times of danger, the testes may be held up in the lower part of the inguinal canal.

The Descent of the Testes

During early foetal life each testis develops close to the roof of the abdominal cavity (fig 10) and close to the corresponding kidney on either side. Later, the testes migrate from this primitive position and pass through the inguinal ring in the abdominal muscle and from here into the scrotum. Each testis is preceded by a pouch of the peritoneum which it carries downward through the inguinal canal together with the muscle, already mentioned, in the spermatic cord.

The mechanical factors which cause the testes to migrate are poorly understood. Ligaments attached to the testes and the epididymis may, by progressive shortening, help to guide the testes and epididymis in their descent through the abdominal wall. Increased intra-abdominal pressure may also play a part. Both testes are usually present in the scrotum or in the inguinal canal at the birth of the foal, but it is probable that in some individuals they can pass freely through the inguinal canal for weeks or even months after the foal is born, provided the inguinal ring remains open. In quite a large number of Thoroughbreds one testis may be small and held in the inguinal canal until the colt is four or five years old. In other cases, one or both testes may be retained permanently in the abdominal cavity, a condition known as cryptorchidism, and the horse is often called a 'rig'. The retained testis is then small, soft, flabby and not capable of producing sperm. One testis may not descend or 'drop' into the scrotum until three, four or even five years of age. In these cases the 'missing' testis can usually be felt in the inguinal canal.

The Accessory Glands

The accessory glands consist of the seminal vesicles, prostate and bulbo-urethral glands (fig 10). They contribute the fluid and substances that, with the spermatozoa, form the semen. The seminal vesicles are elongated sacs lying on each side of the dorsal surface of the bladder. They are about 8in (20cm) long and possess a duct through which their secretions are discharged into the ductus deferens.

The prostate is a lobulated gland which lies on the neck of the bladder and the beginning of the urethra. It consists of two lateral lobes and a connecting bridge. There are two bulbo-urethral glands lying on either

(above) *The terminal part of the urethra is seen protruding from the end of the penis surrounded by the fossa in which there is a sinus containing smegma;* (below) *the sinus has been exposed by drawing the urethra to one side*

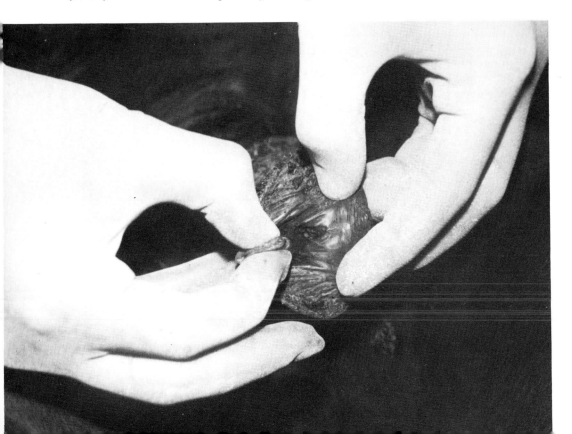

side of the urethra. They are oval and measure about 2in (4cm) long and about 1in (2.5cm) wide.

The Penis

The penis, the male organ of copulation, is composed of spongy, erectile tissue. It is cylindrical in form but compressed laterally. The urethral process and its fossa (fig 12) have become important areas in testing for the presence of venereal microbes, such as *Klebsiella* and the microbe causing contagious equine metritis. Injuries to the organ can cause haemorrhage from the blood vessels, painful swellings and deformation.

The Prepuce

The prepuce is a double invagination of the skin which covers the free portion of the penis when not erect (fig 12). The presence of glands producing smegma, which has a strong unpleasant odour and often accumulates in considerable quantities, is of some practical importance because it may harbour microbes causing venereal disease.

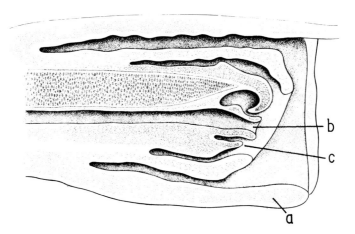

Fig 12 The end of the penis showing the sheath (a), the urethra (b), and the urethral fossa (c)

SEMEN

Semen is a greyish or milky-white fluid containing spermatozoa and seminal plasma. A mature stallion ejaculates around 16fl oz (40 to 80ml) in total volume, although a varying proportion of this may be gel that contains no spermatozoa. Each millilitre (one fifth of a teaspoon) of the sperm-rich portion contains 50 to 100 million sperm. Thus an ejaculate contains some 2,000 to 8,000 million sperm.

Spermatozoon (plural: spermatozoa)

The spermatozoon is the male sex cell (gamete). It is a highly motile cell capable of swimming relatively long distances in the female genital tract, seeking the egg and fertilising it by penetrating into its nucleus and discharging its own nuclear material. The sperm is about one-hundredth the size of the egg, which in turn is no more than the size of a small grain of sand.

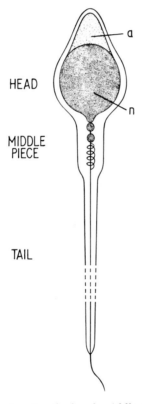

Fig 13 A spermatozoon showing the head, middle piece and tail: the head consists of the acrosome (a) and the nucleus (n) in which the inherited material (chromosomes) is contained; the tail is about ten times the length of the head and a sperm is but a fraction of the size of a grain of sand

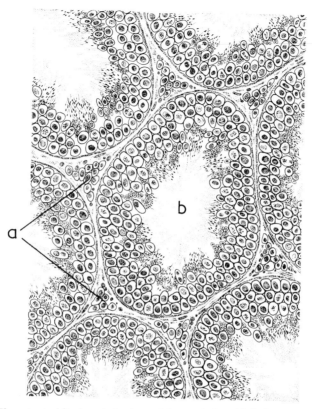

Fig 14 The minute blind-ended tubes of the testis in which the sperm are formed from larger, primitive cells that divide repeatedly in the process of multiplication. Between the tubes are the sertoli cells (a) that secrete the hormone testosterone; the sperm are released into the centre of the tube (b) and from there find their way to the epididymis

A sperm is composed of a head, middle piece and tail. The head contains the nucleus and is capped by a shield (acrosome). The nucleus contains the chromosomes or genetic material. There are sixty-four chromosomes in the horse, made up of thirty-two pairs, each member of a pair being contributed by the male and the female. Thus the sperm contains half the total number of chromosomes normally present in each cell of the body, apart from the red blood cells which have no nuclear material.

The sperm are formed by the multiplication of cells which line the coiled tubes in the testis (fig 14). When multiplying, the number of chromosomes contained in the nucleus is reduced by half, and the mature sperm therefore contain the haploid or half the number of chromosomes present in body cells, ie one-half x sixty-four = thirty-two. After fertilisation has occurred, the thirty-two chromosomes of the egg pair with the thirty-two of the sperm and the body cell number is restored to sixty-four, ie the diploid condition (fig 15).

Seminal Plasma

The seminal plasma is formed by the accessory glands (see page 106). Some ejaculates may contain a gelatinous substance referred to as 'gel'. The gel may comprise up to a fifth of the total ejaculate and it is produced by the seminal vesicles. The quantity of gel may vary with the time interval between ejaculations because it appears that the vesicles are slower to replenish their secretions than are the other accessory glands. More gel is usually ejaculated in the summer than in the winter months; and the quantity varies with the individual in any month of the year. The function of the gel is unknown, but it appears not to bear any relation to a stallion's fertility.

The seminal plasma acts as a vehicle in which the sperm are carried into the uterus. Once inside the uterus the sperm separate from the plasma as they make their way towards the Fallopian tubes. It has been suggested that there are substances in the plasma which kill the sperm after a certain time to prevent them ageing. If the sperm are allowed to age there is a greater chance of foetal defects developing after fertilisation. It may be that some subfertile stallions suffer from an excess of these spermicidal substances and the length of sperm-life is reduced below a minimal level for conception.

Seminal plasma contains substances listed in Table A (see below). The fructose content, thought to be important in the semen of many species, is quite scarce in stallions. Why this should be is not known. The sulfhydryl content of plasma is present in highest concentration in the last portion of the ejaculate, often called the 'tail-end sample'. It is known to be toxic to sperm, and some workers have suggested that its presence may be correlated with fertility, but there is no evidence for this.

Each of the constituents named in Table A can be traced to one or other of the accessory glands. For example, citric acid is derived mainly from the seminal vesicles. The constituents may also appear in different concentrations in sequential portions of the ejaculate. We have already seen that the sulfhydryl occurs in the tail-end of the sample. Ejaculation is usually initiated by the delivery of a watery 'pre-sperm fraction', followed by a milky non-viscous sperm-rich fraction, then a highly

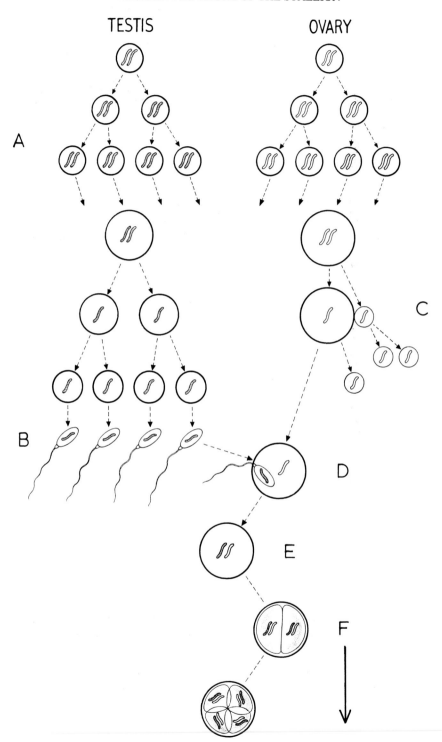

TESTIS

OVARY

A

B

C

D

E

F

viscous and gelatinous post-sperm fraction and lastly, as the stallion dismounts, some fluid is voided from the penis as the tail-end sample.

The group of workers at Krakow, Poland, led by Professor Bielanski, has devised a method of studying the various portions of semen by using an open-ended artificial vagina. This allows the collection, for laboratory study, of each portion separately.

TABLE A Semen and seminal plasma

Average volume of semen – gel-free	50ml
Number of sperm	50–100 million per ml
Total number of sperm in ejaculate	2–5 billion (thousand million)
pH	7.330 pH units

Seminal Plasma	
Specific gravity	1.012
Ergothioneine*	7.6mg/100ml
Citric acid*	26mg/100ml
Fructose*	15mg/100ml
Phosphorus*	17mg/100ml
Lactic acid*	12mg/100ml
Urea*	3mg/100ml

*=average values

Fig 15 One pair of the 32 pairs present in the body cells is shown to illustrate the way in which the chromosome numbers of the body cells (A) are reduced by half during production of the sperm (B) and egg (C) in the testes and ovaries respectively. The full complement of chromosomes is restored at fertilisation (D). The new individual starts as a single cell (E) with half the chromosomes from the sire and half from the dam; this cell then divides and each cell contains the normal number of chromosomes in the body cells (F)

SEXUAL BEHAVIOUR

Developing Patterns of Sexual Behaviour

The attitude of the male to the female horse in the natural state is one of dominance within a small group or harem of five to fifteen members. The social bonding of members of a group to one another and to the stallion is permanent and not related to geographical distribution. In other words, the stallion roams with the harem into areas where they can find food. The stallion, as with mares, is sexually active during the late spring and summer. However, even during the sexually active period, the stallion seldom approaches his mares in an aggressive fashion; rather, he intuitively recognises oestrus using his sense of sight and, more importantly, his sense of smell and taste. We do not know the factors which cause the stallion to mate at any particular time with any member of the harem that is in oestrus. However, individuals may develop a pattern of sexual approach so that they copulate in the early morning and late evening. It does not seem that the male horse necessarily mates a mare in oestrus several times in each day or, indeed, on each day during a heat period.

We must not, however, dwell on the behaviour of stallions in the natural state because this hardly applies to any situation, except in the more remote parts of the world. Breeding in hand is by far the most common practice and even feral (semi-wild) horses, allowed to roam in herds in the Camargue, the New Forest, Exmoor or Dartmoor, are largely controlled by man; their members re-arranged arbitrarily for human convenience, such as for riding and breeding for type.

In practice, by the time the potential stallion arrives on the studfarm his behaviour is already modified. In the natural state, the young colt is weaned at about eleven months of age, when its dam prepares to give birth to her next foal. The colt remains in the herd until it is perhaps two or three years of age, when it is driven off by the stallion of the harem. It joins a bachelor group of similarly treated youngsters, but finally wanders away in search of sexual satisfaction. It soon proceeds to create its own harem, composed of mares that have become separated from their original group, or maiden fillies who have separated from their original harem.

In strong contrast, the working or racing horse is reared under greatly different circumstances. From an early age the colt enters the system of re-arrangement of herd membership. This is imposed by the necessity of moving mares from one studfarm to another and, within each studfarm, from one paddock to another, according to mating plans and managerial needs for modern stud husbandry. Thus, the social bonds of the herd are continually disrupted. Periods of isolation in loose boxes are additional disruptive elements in the life of the colt, continuing beyond weaning, which is in itself an abrupt and disturbing process.

When the colt arrives at the studfarm to take up stallion duties, he has thus been conditioned to social and sexual responses which are very different from those which lead naturally to life within a harem or group of mares. In fact, sexual activity has been discouraged and reduced to minimal proportions for man's convenience. Behavioural patterns have been assiduously developed, by training for racing or other artificial pastimes, and the colt conditioned to a life relatively isolated from physical and sexual contact with other horses. The stallion is selected for his working prowess and his ability to conform to man's managerial requirements. It is important for us to appreciate the natural background of a colt's sexuality to appreciate the best methods of management, within the limits imposed by commercial and other factors. It seems inevitable that we should breed horses in hand, keeping the stallion separated from his mares, having sexual contact only at times of mating and, possibly, at those of teasing. But, by attempting to understand the urges of a stallion to behave in a manner natural to his basic psychological make-up, we may be able to avoid the more disruptive effects of artificial measures.

Patterns of Mating Behaviour

The sight, smell and taste of the mare in oestrus arouses the stallion through nervous and hormonal pathways. There are essentially six components of sexual behaviour, namely the approach, penile erection, mounting, intromission, pelvic thrusts, ejaculation and dismounting.

The approach is composed of a visual appraisal of the mare followed by physical contact of nuzzling around the flanks and hind quarters. This preliminary courtship develops the libido necessary for the sexual act. It is in essence a low-key pattern of behaviour, typified by gentle biting and the occasional Flehmen's posture. Stallions that are bred in hand, and teasers, who have rare opportunities for mating, may develop an aggressive style of courtship with quite vigorous and even savage biting. The length of courtship varies with the individual, but in the case of breeding in hand it is regulated by the attendant who watches for the horse to 'draw' (unsheath the penis).

The act of mounting leaves the horse in the most vulnerable position, should the mare not be in oestrus. Some stallions will make preliminary up-and-down movements with their chest against the mare's hind quarters

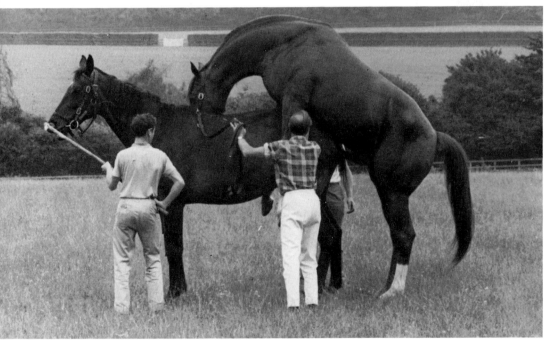

Mating behaviour can be roughly divided into the components of approach
(opposite, top), *erection* (opposite, centre), *mounting and intromission*
(opposite, bottom), *pelvic thrusts and ejaculation* (top) *and dismounting*
(bottom)

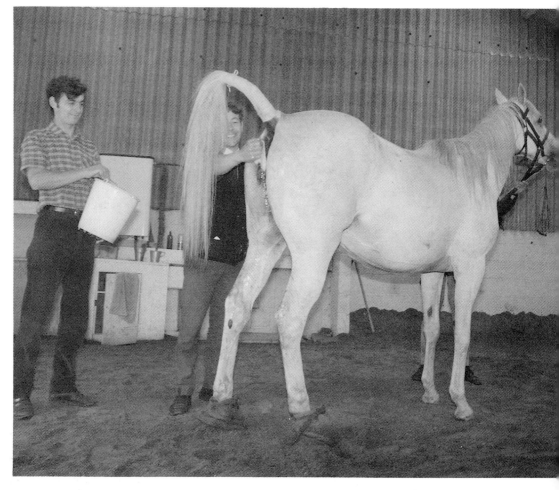

Washing the mare's perineum with water in preparation for mating

before actually rising to mount. This nudging may be seen as a preliminary teasing manoeuvre to make the mare show any tendency to
aggressive behaviour prior to the act of mounting. A kick, as far as the
stallion is concerned, is likely to do less damage while he has four feet on
the ground than if he is in the rearing position.

Mounting usually takes place at a slight angle to one side of the mid
line. The attendant (stallion man) holding the stallion stands on the left
side of the mare. The act of entering the mare (intromission) may be
affected by the size of the glans penis (rose). The glans increases in size
after the stallion has entered the mare, but in some cases may do so prior

to intromission. This may be due to the horse being restrained too long before being allowed to mount. A further influence is the shape and conformation of the mare's perineum and vulva. If this is sloping and/or has had a Caslick operation previously performed on it, intromission may be difficult.

Once inside the mare the horse delivers a series of pelvic thrusts, maintaining his grip and balance with his forelimbs and, in many cases, by grasping the mare around the withers or lower part of the neck with his teeth. Aggressive behaviour with the teeth may cause injury, and a special leather roller or pad may be placed around the neck of the mare for the horse to grip with his teeth.

Ejaculation is usually accompanied by some sign such as flagging of the tail. In some individuals, however, it may be difficult to assess that ejaculation has taken place without feeling for the pulse wave passing along the urethra on the lower border of the penis. Some horses resent even this minimal amount of handling necessary to palpate the seminal pulse wave. In these cases, those present have to rely on visual signs to determine whether or not ejaculation has occurred. A knowledge of the horse is important because the casual observer may make a misdiagnosis. The importance of knowing whether or not a horse has ejaculated cannot be overemphasised (see page 124).

Feeling for the pulse wave along the urethra at the time of ejaculation

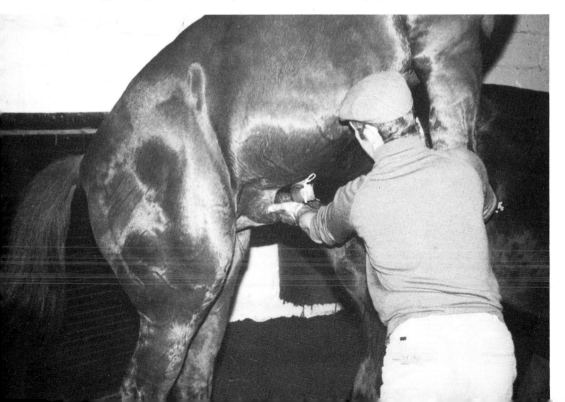

The time of dismounting varies with the individual stallion and, to some extent, with the mare and with environmental factors. A restless mare, noise or movement on the part of attendants present in the covering yard may shorten the time during which the horse remains in position following ejaculation.

MANAGEMENT OF THE STALLION AND MARE AT MATING

The stallion occupies a privileged place on the studfarm; there are many mares but few stallions. The male horse has to be treated circumspectly. Most are reasonably tractable, but there is no knowing when they may show flashes of aggressive behaviour, based on their natural inheritance, to dominate male competitors that may appear to threaten their supremacy. In the distant past, Thoroughbred colts in training and stallions were often vicious, and as they grew older they became especially dangerous. Nowadays they are on the whole more tractable and easier to handle. Nevertheless the records of recent times show that there have been a number of incidents in which men have been savaged by stallions sometimes acting quite out of character. No one should, therefore, attempt to handle a male horse over the age of four years without having had some experience and without always being on their guard against some unexpected attack. The awe in which stallions are held is a correct and prudent attitude.

Mating

The approach of management to mating mares with stallions must be based on the principle that is the binding theme of this book: namely, that we cannot improve on natural functions, but must aim to avoid harmful actions in our attempts to fashion nature for our own ends. The stallion and mare running free have no problem in mating; it is only when they are led in hand that there is a need for restraint, to avoid possible injury and to conduct the process within a given period of time. The mare is led into the covering yard and introduced to the stallion across the trying board. If she shows sufficient signs of oestrus she is then 'prepared' in diverse ways, according to the particular practice of the studfarm. In most cases, the top of the tail is bandaged and the buttocks and perineum are washed with soap and water, or with water and disinfectant. It should be emphasised, however, that both soap and disinfectant may have a contraceptive action.

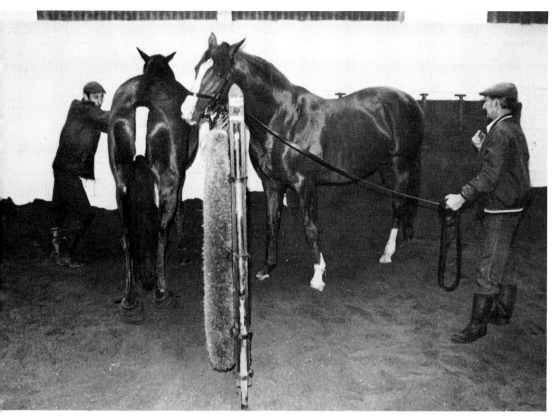

Mare and stallion may come together for the first time across the teasing board in the covering yard

However, provided excess is not left on the coat and skin of the mare there is probably little risk on this score. A more serious problem might arise if the stallion objects to the smell and taste of the preparation used.

Felt boots are placed on the mare's hind legs and, after positioning in a suitable area of the yard, a twitch may be applied to the mare's nose. A front leg may, if this is thought necessary, be held or strapped in the flexed position. The mare is now ready for mating. It used to be common practice for hobbles to be applied in addition to the precautions already mentioned, but in the UK this practice has been discontinued, although it is still practised in some Southern Hemisphere countries.

Most of the precautions mentioned are quite unnecessary if the mare is strongly in oestrus. In fact, the necessity for applying them decreases in direct proportion to the strength of oestrus. Because mares vary in this regard, it is usual for some precautions to be taken in all cases. Nor should we expect mares to submit to mating, in these unnatural circumstances, in the same manner that they would if running free with the stallion. Within

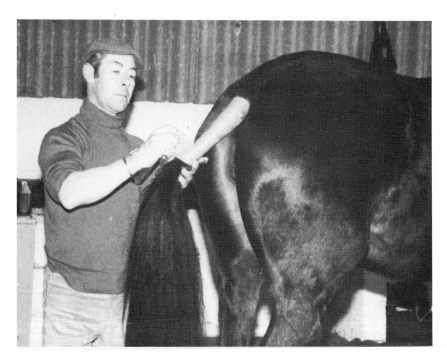

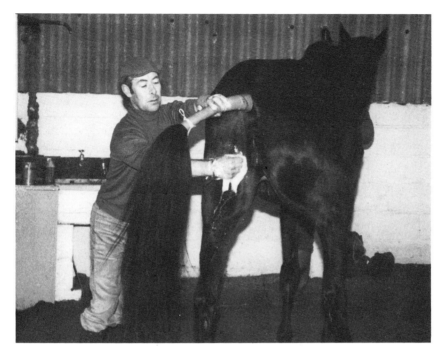

(opposite and above) *Preparing the mare for mating*

the environment of modern stud management, the sex life of the mare can hardly be classed as normal or natural; and contact with the male horse in the covering yard can hardly be regarded as natural courtship. Maiden mares and those with relatively low libido may require some restraint if they are not to be of danger to the horse. Apart from the restraints already mentioned, it may be helpful to put blindfolds on mares if they are frightened by the approach of the horse from behind; even tranquillisers may be necessary to facilitate the process of mating in certain instances.

Once the mare has been suitably prepared, the stallion is led to a position behind and to one side. He is allowed to sniff and touch the mare's flanks and hind quarters, being restrained from any attempts to bite or to mount before he has achieved an erection. When he is judged ready to mount, the attendant or stallion man pulls the leading rein over the mare's back. With an experienced horse this acts as a signal to mount. Intromission may be effected with ease or, if the conformation of the perineum is deep-set and narrow, with difficulty. And it may be a problem for a horse to enter a mare that has a stitched (Caslicked) vulva. The man holding the stallion or the mare's tail may, in these cases, assist by

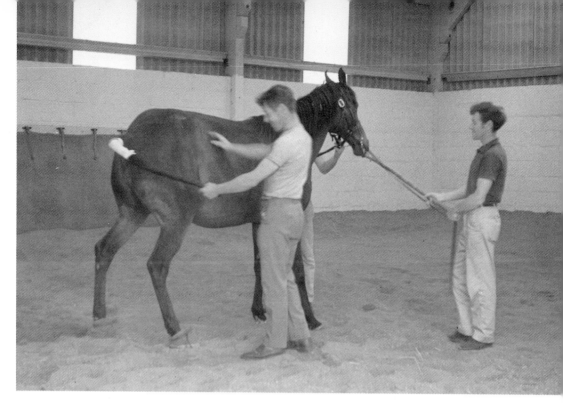

Positioning the mare for mating

grasping and directing the penis. This action is routine practice on some studfarms. However, many horses resent the procedure and may dismount immediately.

The pelvic thrusts leading to ejaculation must be carefully watched for evidence that ejaculation has in fact occurred before the horse dismounts. Ejaculation is usually accompanied by rhythmic movements of the tail, but as already suggested the seminal pulse is a more reliable guide.

The ejaculatory process may be delayed or made difficult by a dispro-portion in size between the mare and the stallion. A short-backed horse and a tall, long-backed mare, for example, may make it difficult for the stallion to work. The effect of this disproportion may be overcome to some extent by digging a pit in which the mare's hind legs may be posi-tioned. Other factors which may prevent ejaculation are low libido or a painful lesion in the stallion's back or in the penis (see page 263).

Libido

Libido is the sexual drive, and its strength varies between stallions, the season of the year and the number of services in the day or week. Dr Bill Pickett and his colleagues at Colorado State University Experimental Station, Fort Collins, measured the reaction time of mating in a number of ways – for example, in terms of the minutes required for the stallion to mount the mare and begin copulatory movements. The mean reaction time was 3.5 minutes for first ejaculates and 3.7 minutes for second ejaculates, occurring one hour after the first. However, the season had considerable influence, and in the case of first ejaculates the interval between the introduction of the horse to the mare and ejaculation varied from 1 minute in May to 9 minutes in December. Whatever the value of the actual timing, it is clear from these results that reaction time decreased as the breeding season approached. The group at Colorado found that, on average, 1.8 mounts were required before ejaculation occurred. But, as might be expected, this measurement was also influenced by the season.

The most important lesson that may be drawn is that those individuals with a strong sexual drive may mate two or three times a day at any season of the year, but many individuals may lose their drive outside the natural breeding season of spring and summer; and, more importantly, may be affected by over-use.

The average number of sperm of gel-free semen is about 100 million per millilitre. However, second ejaculates, that is those collected approximately one hour after the first, contain only about sixty per cent as many sperm as the first ejaculate. Season may also influence the concentration of sperm, following the same pattern as for other measurements, namely better results in the spring and summer than in autumn and winter.

The total number of spermatozoa per ejaculate is about 5,000 million in first ejaculates and 3,000 million in second ejaculates. But there is a marked seasonal variation; stallions examined in January may have half the number of those examined in July. The number of services per day and per week also influences results. Dr Pickett and his co-workers collected semen from a stallion at hourly intervals. The total sperm in each ejaculate decreased from about 60,000 million to 8,000 million over the five-hour study. These figures provide an example of how the use of stallions may affect semen quality. However, quality has to be measured not only by numbers, but by motility and structure, ie normal compared with abnormal forms. Motility is evaluated by examining the sperm under a microscope, using standard techniques and pre-warmed utensils. It is important that the semen is not chilled, as this may in itself injure the sperm and reduce the motility. A drop of semen is placed on a cover slip and inverted over a shallow cavity in a glass slide. This 'hanging' drop is then examined under a microscope which magnifies about 200 times. The sperm are observed showing progressive motility, that is moving in a relatively straight line across the viewing field. The percentage motility in a good sample should be in the region of sixty per cent motile sperm. This number decreases hourly, although after three hours at room temperature some motility should still be present.

Evaluating Semen Quality

Vets and breeders often need to assess the fertility of stallions, in terms of the quality of their semen. This is not a simple matter because of the great variations in quality encountered in fertile and infertile stallions. Thus it is possible to have an individual with apparently poor-quality semen obtaining satisfactory results at stud, whereas another individual with comparatively good semen quality proves to be infertile. This discrepancy, which causes considerable embarrassment at times to those who have to provide a forecast, is due to missing factors about which we have little knowledge at the present time. These 'missing links' have engendered tests to estimate the ability of the sperm to reach the egg in the genital tract of the mare and to fertilise it. In other words, we need a means of measuring the fertilising capacity of the sperm, rather than knowledge of their morphology (form) and numbers, valuable as this information may be.

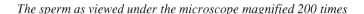

The sperm as viewed under the microscope magnified 200 times

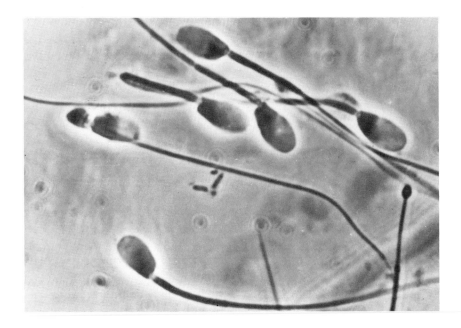

If we assume this to be the case it follows that failure to ripen may be one reason for the failure of apparently normal sperm to have a fertilising capacity. Further, it seems that the ability for a sperm to enter the egg may perhaps depend on the presence of many other sperm, playing a part in the process by their presence as catalysts. Thus, the conditions for successful fertilisation are the presence of ripened spermatozoa in the ampulla of the Fallopian tube, although the minimal number necessary for success is unknown. The ability to reach the oviduct has been shown, by scientists at the University of Wisconsin, to vary significantly with the stallion. This finding suggests that the fertilising capacity of the semen may be related to some energy factor in the sperm and/or a factor that enables the sperm to survive the hostility of the mare's genital tract. None of these influences can be measured by the standard techniques of semen examination.

Current methods of assessing semen quality are by volume, sperm morphology and motility and by the period sperm survive under standard conditions. If these measurements are markedly subnormal, it is probable that the stallion will be subfertile. The important indicator is that there must be a minimal number of normal live sperm in the ejaculate to achieve successful fertilising capacity. This can be better explained by taking an extreme example. If all the sperm have normal forms but are dead at the time of ejaculation, there is no chance of conception occurring; nor if there are one hundred per cent abnormal forms, even though all the sperm are alive and motile. The critical test is the number of live, motile and normal sperm in the ejaculate and this is a sum of (1) volume of the ejaculate, (2) concentration of sperm per millilitre and (3) the percentage of live, normal, motile sperm in the sample. The lower limit of quality has been suggested as 500 million live, normal sperm in the ejaculate.

If we suppose that there is a minimal quality of ejaculate in terms of the number of normal, live, motile sperm, we can appreciate how other factors, such as season and the frequency of ejaculations, are important, relative to conception in a horse with poor semen quality. Drs Pickett and Voss have shown that stallions ejaculating six times a week have a lower volume of gel and sperm concentration than those ejaculating once a week. In these authors' opinion, one of the most important contributions to infertility is the over-use of stallions. On Thoroughbred studfarms it is usual for stallions, during the breeding season, to mate two or three times a day, ie fourteen to twenty-one times a week. The number of services in a breeding season of twenty weeks for a stallion with fifty mares may be expected to involve one hundred to one hundred and fifty matings, depending on the fertility of the mares and on managerial control, including veterinary examinations. There is, unfortunately, no reliable evidence of how this repeated use of stallions may affect an individual's semen quality. It would be an interesting and practical step to make routine semen examination of Thoroughbred stallions at regular intervals during the breeding season. If we did this we would probably find that some individuals were more affected than others by over-use. The most important practical aspect is whether or not the way in which a horse is managed may reduce the number of live, normal sperm, capable of fertilising, being delivered into the uterus at the time of ejaculation. If a stallion has poor-quality semen and he is over-used, the result is a decrease in the number of live, normal sperm contained in each ejaculate. The lowest limit may be breached and the ejaculate be of insufficient quality to achieve conception.

It has been shown that sexual stimulation, prior to service, increases the sperm output of bulls. However, this does not seem to apply to stallions, according to studies made at Colorado State University. Teasing appears to increase the volume of gel and gel-free semen and to decrease the concentration of sperm (number per millilitre). The conclusion is that the teasing of a stallion before the mating process is unlikely to achieve a greater number of sperm inseminated through ejaculation at the time of service. In increasing the quantity of gel, threefold in some cases, the process may actually be harmful, although there is no scientific information regarding the effects of gel on the stallion's ability to achieve conception in mares. However, using the stallion as a teaser may have a significant effect on the stallion's reaction time and 'slow' performers may be used in this way prior to mating.

The state of the mare's genital tract must also play a part in determining whether or not a particular ejaculate is capable of achieving fertility. If the uterine environment is particularly hostile to the sperm, because of infection or some other factor, the fertilising capacity of the ejaculate would need to be that much greater for the sperm to survive.

Fertility and Subfertility

A fertile stallion is one which delivers semen of the highest quality and is, therefore, least affected by the variables described above; a subfertile stallion, on the other hand, is one that may be seriously affected by adverse conditions such as over-use and mating with subfertile or 'difficult' mares. A sterile stallion is one where these handicaps are absolute and the conception rate in the mares mated is zero. The reader should appreciate that the definition of these terms is purely arbitrary and, further, that the level at which the horse may be described as infertile or subfertile is a matter of semantics rather than science. In the case of insurance against infertility, stallions may be insured by underwriters against failing to achieve sixty-per-cent levels of conception over the breeding season. This level recognises that most stallions achieve rather better results among their harem of forty to fifty mares. However, one has to be careful in discussing these figures because it is first necessary to define the endpoint on which the stallion's fertility is judged. For example, pregnancy diagnosis at, say, one-third of the way through pregnancy or at full term is a rather different endpoint from a diagnosis made at twenty days. There is a natural wastage of at least 10 per cent from twenty to sixty days of pregnancy and of a further 10 per cent between sixty days and full term. Thus, the fertility of a stallion estimated on twenty-one day pregnancy examinations may be at least 15 per cent more than if it were estimated at, say, an examination made at five months of pregnancy. Further, the results depend on the number of mares served during the stud season; the higher the number, over about thirty days, the less likely that a hundred per cent result will be obtained. In addition the fertility of the mares and the way in which the horse is managed affect the conception rate, quite apart from the basic fertility of the stallion.

Hormonal Control

The output of sperm and the sexual behaviour of stallions are to some extent controlled by hormones secreted by the cells of the anterior pituitary gland and by the testes.

The pituitary produces luteinising hormone (LH) and follicle-stimulating hormone (FSH). These hormones are known as gonadotrophic hormones because they stimulate the gonads, in this case the testes and, in the mare, the ovaries (see page 71). There is a similar relationship between the pituitary and the brain in the stallion as there is in the mare. The cells of the anterior pituitary are themselves under the control of the hormone GNRH (gonadotrophin-releasing hormone) produced by special cells in the brain.

The hormone LH affects the *Leydig* (interstitial) cells of the testes, causing them to produce testosterone. There is usually a rapid response to the secretion of LH, and testosterone levels rise in the blood stream, resulting in a masculinising influence on many tissues in the body. The testis itself is affected by the testosterone although there is some disagreement among scientists as to the extent that it affects sperm production. The main influence on spermatogenesis (sperm-forming activity) is FSH. This hormone stimulates the sperm-producing cells to produce new spermatozoa. However, it should be emphasised that, although the roles of LH and FSH are separate within the testis, these two hormones do not function independently, but in parallel, and promote each other.

Rising levels of testosterone in the blood stream tend to depress the output of GNRH, LH and FSH; but as levels of testosterone subsequently fall so the levels of GNRH and the anterior pituitary hormones are once again secreted to stimulate the further production of testosterone. Thus the hormonal system ensures that testosterone is constantly present at low levels to perform its masculinising effect, but peaks occur related to sexual activity. The nervous pathways from the brain associated with sight, smell and taste ensure that GNRH and, consequently, LH and FSH are released when the male horse is stimulated by the female; consequently testosterone levels increase. Similarly, there is a distinct seasonal variation in the levels of pituitary and testosterone hormones mediated through the seasonal influence of light. We have already discussed the way in which libido and semen quality is improved during the natural breeding season; and this improvement depends on the pathway of brain–GNRH–pituitary–testes.

ARTIFICIAL INSEMINATION (see also chapter 15)

In many countries, the practice of artificial insemination (AI) is a well-established technique. However, it is not at present allowed to be used on Thoroughbreds. This is a largely unjustifiable restriction based on traditional fears that the breed might in some way be damaged by malpractices and over-use. This is not the place to argue in detail the case for or against AI, but a summary of the technique and its advantages and disadvantages is appropriate.

Technique

The technique of AI is a simple one based on the use of an artificial vagina (AV). This consists of an outer rigid casing with a rubber internal lining open at one end for entry of the stallion's penis and at the other for collecting the ejaculated semen. There is a space between the metal and rubber linings for water, which is maintained at blood temperature and at suitable pressure. The inside of the rubber lining is lubricated by Vaseline or special jelly.

The apparatus is held by the operator to one side of the mare as the stallion mounts. He guides the horse's penis into the AV, maintaining it at an appropriate angle to the horse. After ejaculation, the sample collects at the farther end in a container attached for the purpose.

It is essential that all surfaces with which the semen comes into contact be warmed to at least blood temperature. Rough treatment can affect the sperm adversely and thus reduce the sample quality. The ejaculate is then separated into gel and gel-free components. The gel-free part may be used in one of the following ways:

1 Inseminated immediately into a recipient mare.

2 Stored at temperatures above freezing point and transported to a recipient mare at some distance from collection or on other premises. For this purpose the semen may be incorporated into a special mixture containing substances which protect the sperm. This fluid is sometimes called semen extender; it usually contains such constituents as skimmed milk and antibiotics. It is inadvisable to store semen for more than two or three hours before use.

3 Divided into several portions depending on the volume and quality of the sample. An ejaculate of good-quality semen of average size (16fl oz, 40ml) may be split ten to twenty times.

4 Frozen by special methods to protect it, and used weeks, months or even years later in single or multiple doses.

The split ejaculate is used on Standardbred farms to enable a large number of mares to be inseminated each day from one ejaculate. This practice reduces the number of ejaculates to a very small number relative to the number of mares inseminated. It also enables management to have mares inseminated daily throughout the later period of oestrus; thus ensuring that some semen is inseminated close to the time of ovulation. This may make a significant contribution to a mare that is subfertile and in whose genital tract the sperm lives for a relatively short period, due to the exceptionally hostile environment of an infected or diseased tract.

However, there are risks in splitting a semen sample if a stallion has poor-quality semen. In fact, most Thoroughbred stallions seem to have a rather high proportion of dead or abnormal sperm and in this breed, at least, it might be foolish to divide the semen. Using the whole ejaculate per inseminate provides a useful basis to overcome the fears of tradition-alists and others who believe that using AI would mean the over-use of certain blood lines at the expense of others. No more mares could be served than achieved by natural coitus.

Control

Veterinarians are anxious that AI should be allowed under strictly controlled conditions in circumstances where there is a veterinary necessity. The indications include its use on mares that are suspected of harbouring such venereal infections as contagious equine metritis organisms, *Klebsiella* and *Pseudomonas*. Mares that have recently been infected or those that have had contact in previous years may be safely inseminated by AI, but could pose a considerable risk if served naturally. AI is, therefore, a very useful means of prophylaxis. Of course, it may not be advisable to inseminate some infected individuals until suitably treated. This is a veterinary decision which must be taken by those responsible and with knowledge of the particular circumstances of the case.

Further, the susceptible mare (see page 254) is one that may succumb to the coital challenge of infection by whatever microbes might happen to be on the horse's penis and inseminated into the uterus at the time of coitus. These susceptible mares would benefit by the use of AI by removing the infective challenge and, therefore, providing a greater chance of conception.

The use of AI must be properly controlled in Thoroughbreds. A sensible approach would be:

1 To allow its use when certified by one or two veterinarians that there is a bona fide reason. The technique should be applied only by veterinarians, so as to ensure that the protocol can be controlled through a central body, in this case the governing body of veterinarians. The certification of procedure can then be backed by disciplinary action should the need arise. Vets have a professional commitment to accuracy when certifying the facts of any situation. They are thus admirably suited to a system designed to prevent fraud and the malpractice of artificial insemination procedures.

2 To ensure that each ejaculate should be inseminated into one mare only, and not split for use on any other mare. No stallion could, therefore, be mated with more mares than he could reasonably serve by natural means. In fact, the number of services per conception would remain the same as for natural service. There would thus be no reason to fear that one horse would receive an inordinate number of mares in any breeding season; nor that the genetic material of any outstanding horse would be

inseminated more widely among the breed than under existing conditions where natural coitus is practised.

AI could provide considerable benefit in the control of disease; its use, based on the two principles of certification and of 'one ejaculate one mare', would ensure that the technique had no detrimental effects on the integrity of stud books. The procedure is obviously limited by cost and by the fact that some stallions would not tolerate the procedure. Further, if the technique were to be limited to cases where there was a veterinary necessity, the number of instances in which AI would be used would remain relatively small.

There are a number of subsidiary control measures that might be used to police AI, including blood typing tests of progeny, a compulsory registration of horses and mares used in the procedure, and a proviso that the produce of any mating illegally or irregularly obtained by AI should be banned not only from the stud book but from the racecourse.

The rules outlined here are, of course, a political compromise due to the attitude of the Thoroughbred stud book authorities. The reader should not be led to believe that there is something intrinsically wrong with AI. It has been used extensively in Poland, for example, to obtain better fertility among mares than was previously experienced under natural breeding conditions.

5
PREGNANCY

THE WAY OF MAMMALS

Pregnancy is a period in which the new individual (foetus) achieves sufficient maturity to survive as an independent being. It does not become
entirely independent until it is weaned, some weeks or months after birth.
The problem of protecting and nourishing the fertilised egg, until it has
developed to a mature state, is one faced by all multicellular animal life.
Mammals nurture their young within the female on the principle that the
maternal being supplies vital substances to the developing individual
within its womb, receiving back waste material which would otherwise be
poisonous.

The Placenta

The organ developed for this exchange is the placenta, which is attached
to the wall of the uterus (fig 16) and connected to the developing foetus by
the umbilical cord. The cord contains vessels through which blood circulates to and from the foetus. In the placenta the foetal blood stream comes
into close contact, although never mixing, with the maternal blood stream
flowing through the uterine wall. The manner in which the placenta is
attached to the uterine wall differs according to the species.

Foetal Environment

Let us now consider some of the other necessities of foetal life, and how
mammals in general provide for their young during pregnancy. The
placenta takes on the functions which in the born animal (ie during life
after birth) we recognise as feeding, breathing, urinating and defecating.
Life before birth presents problems to the foetus similar to those experienced by the born individual. For instance, both have to survive the
external environment of temperature and gravity and physical factors
including pressure, friction, trauma. The foetus might, if it could talk,
explain as Shakespeare's Shylock, 'If you prick us, do we not bleed? If
you tickle us, do we not laugh? If you poison us, do we not die?'
    The mammalian foetus lives, as it were, in a pond buried deep within

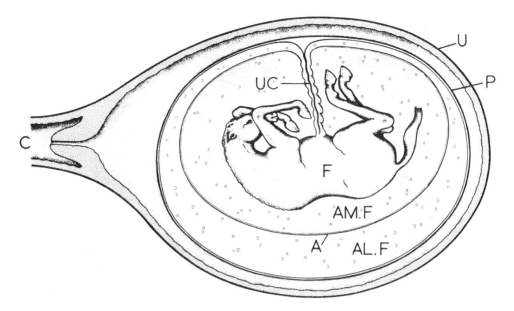

Fig 16 The foetus (F) obtains nourishment from the placenta (P) attached to the uterine wall (U); the umbilical cord (UC) connects the placenta to the foetus, which is surrounded by the amnion (A), a membrane containing the amniotic fluid (AM.F); the placenta contains another fluid – allantoic fluid (AL.F). The uterine contents are closed to the outside during pregnancy by the cervix (C)

the earth. Fluids surround and cushion it against the effects of gravity and untoward movements, just as we might be protected when floating in water. The effects of light, sound and outside movement are reduced. The foetus is further protected by the wall of the uterus and the surrounding organs contained within the mother's abdomen. The foetus cannot lose heat or water except through its surroundings; and the mother, thereby, controls the foetal heat and water balance. Microbes are unable to enter the foetal body except by crossing the placenta. Thus this organ forms a natural barrier, breached only in rare circumstances.

The Foetal 'Parasite'

The developing foetus is a new individual, albeit a mixture of its parents. However, it represents a separate genetic entity from its mother, and the foetal and maternal tissues are foreign to one another. The relationship between the mammalian female and its foetus is similar to that of host and

135

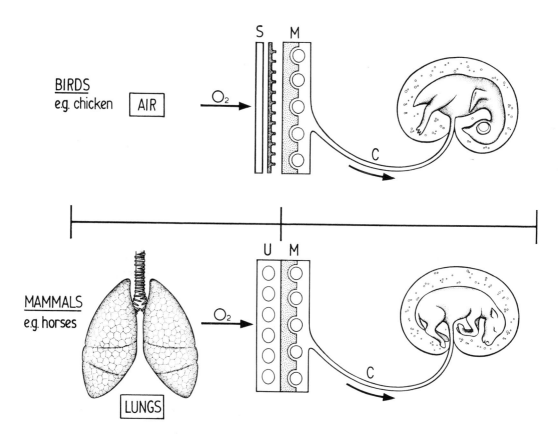

Fig 17 The difference between the way birds and mammals supply their young with the vital gas, oxygen, during embryonic development. In the chicken, oxygen passes through the shell (S) into the membrane (M) surrounding the chick; here it enters the blood stream and is carried in the cord (C) to the developing embryo. In the horse the mare takes oxygen into her lungs and this is carried in her blood stream to the uterus (U) where it crosses into the membrane (M) or placenta surrounding the foal; in the placenta, oxygen enters the foetal blood stream and is carried to the developing foal in the cord (from Steven, D.H. and Samuel, C.A., 'Anatomy of Placental Transfer' in *Placental Transfer* eds Chamberlain, G.V.P. and Wilkinson, A.W., 1979)

parasite. A parasite normally provokes from the host an immunological response which rejects foreign material. This process is encountered whenever foreign matter enters mammalian body tissues. For example, a skin graft or kidney transplant is rejected by the recipient because the donor tissues are recognised as foreign, in the same way as a parasite or microbe. In the case of pregnancy, we do not know for sure how the immune response of the mother is enabled to accept the foetal 'parasite'. However, the survival of the foetus obviously depends on the fact that some such mechanism exists.

Maintenance

It is necessary for the foetus to be retained in the uterus during the entire length of pregnancy. The uterus has ideal properties for the purpose because its walls are able to stretch as the foetus increases in size. The exit, through the eventual birth canal, is kept sealed by the cervix. Almost all the means of maintenance are concerned with hormonal control. The chief hormone of pregnancy is progesterone, but oestrogens, prosta-glandins, cortisone, prolactin and gonadatrophins have roles in regulating cellular activity, promoting a union at the placental-uterine juncture and influencing, directly and indirectly, the growth and development of the foetus. The foetus has a complete complement of its own hormones, and the degree to which maternal and foetal hormones play specific roles varies from one species to another.

THE WAY OF THE HORSE FAMILY

We must now turn to our main inquiry, namely how does the horse achieve the objectives of pregnancy through its own special type of placenta and maternal/foetal interactions?

Fertilising the Egg

The egg enters the oviduct down its funnel-shaped opening formed by the fimbriae. It travels to the isthmus, a small dilatation, where it is met by a swarm of spermatozoa, one of which penetrates the outer membrane of the egg and injects its genetic material. Thus the fertilised egg contains 32 + 32 chromosomes.

Once fertilised, the egg immediately becomes resistant to the entry of any other sperm. The part played by the thousands and thousands of other sperm in the process of fertilisation is unknown, though their presence seems to be necessary for the entry of the successful spermatozoon. Further, it seems probable that the sperm undergo ripening (capacitation) by which their acrosomes (page 110) are made capable of penetrating the egg, while they are in the female genital tract. It is not known, in the case of the horse, for how long the sperm have to be present in the genital tract before they ripen, but it is probably a matter of hours.

The egg, if it is not fertilised, 'dies' within about twenty-four hours. In practice, the chances of fertilisation diminish rapidly as soon as the egg enters the oviduct. In practical terms this means that a mare should be served by the stallion *before* and not *after* ovulation. This ensures that the sperm are present in a 'capacitated' state at the site of fertilisation when the egg arrives in the oviduct.

Terminology

It might be helpful to define some of the terms which will be used in this chapter. The *embryo* is the egg as it develops into a horse-like form, and embryology is the study of its development. The term *foetus,* on the other hand, may be applied to the older embryo, say, after sixty days from fertilisation. However, for our purposes, the terms embryo and foetus are interchangeable. *Gestation* is another word for pregnancy, and the age of gestation denotes the number of days from fertilisation of the egg or, in practice, from the last date of service by the stallion. The *foetal membranes* include the placenta, umbilical cord and amnion. There are two *foetal fluids,* namely, amniotic, contained within the amnion, and allantoic, contained within the placenta. The amniotic and placental membranes together comprise the *afterbirth. Full term* implies a length of gestation at which the foetus is fully formed and mature. In a Thoroughbred this is normally between 320 and 360 days and in smaller breeds between 315 and 340 days. The normal gestational length in donkeys is 360 to 380 days, and in hybrid crosses between horses and donkeys (mule or hinny), it is about halfway between that of the two parental species (340 to 360 days).

Foetal Development

The fertilised egg descends in the Fallopian tube as a single cell, dividing into two, then four, eight, sixteen cells, and so on. The mare, relative to other species, is special in that only fertilised eggs pass into the uterus. The unfertilised egg may undergo a small number of divisions but eventually degenerates and remains in the tube. Neither the reason nor the mechanism for this phenomenon is clearly understood. It may be due to differences in weight and surface properties of the fertilised compared with the unfertilised embryo.

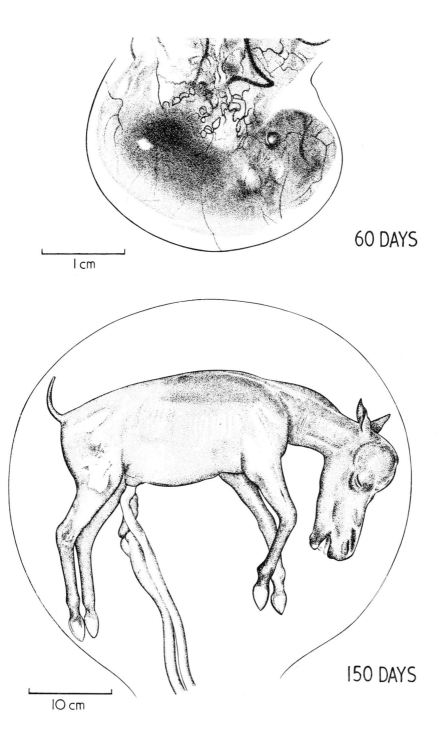

60 DAYS

1 cm

150 DAYS

10 cm

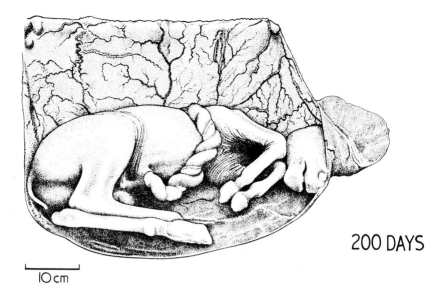

200 DAYS

10 cm

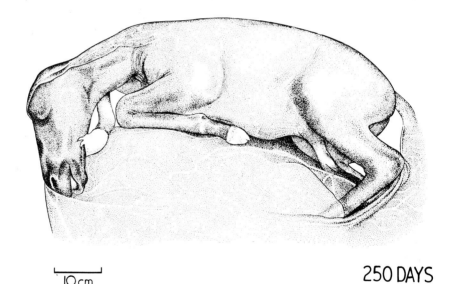

10 cm

250 DAYS

Figs 18 & 19 Four stages of foetal development illustrating the increase in size between a foetus at 60 days to one at 250 days of age; the size of the foetus can be gauged from the scale. None of the foetuses possesses hair except the one at 250 days where hair is growing round the eyes and along the mane, back and tail; by 300 days a coat is present over the entire body

The newly formed embryo reaches the uterus on the sixth day following fertilisation. The cells, as they continue to divide, arrange themselves around a central cavity, and thus begin the two major embryological processes of *differentiation* and *growth*. The ability of cells in the developing embryo to differentiate into specialised layers and tissues is the basis of organ formation and characterises the highly complex organisation of mammalian life. Growth is necessary to achieve the size required for independent survival.

The two processes of differentiation and growth are continuous, but occur at differing rates during gestation. In general, differentiation is most intense during the first 40 days and growth the most obvious between 150 and 250 days. However, in relative terms, the embryo increases enormously in size during early development. For example, at the time of fertilisation it is the size of a grain of sand; by nine days it is 0.1in (2mm) in diameter; and two weeks later about 2in (60mm): a 30-fold increase. The proportions of the foal expressed in weight and length throughout pregnancy are shown in Table B.

By Day 16, small blood vessels are appearing from the amorphous mass of cells composing the embryo. By Day 23, the embryo has a definite form with internal organs and limb buds sprouting from either end. By Day 36, the equine characteristics of a horse are readily apparent, in the shape and conformation of the embryo.

Although differentiation is dramatic in the early stages of pregnancy – from a roughly spherical one-cell fertilised egg to a miniature horse is a truly remarkable change – it proceeds at a much more leisurely pace thereafter. For example, although skin is formed in the very early stages, hair only appears from about Day 220, starting to grow round the eyes and along the mane (Table B). The coat covers the entire body only by about Day 300. The differentiation thus follows a logical pattern; the vital organs, including the heart, liver and kidneys, necessary for foetal life develop first; and hair, essential for keeping warm and existing in the outside world, develops last. Within the womb the body temperature of the foal is maintained by the mare.

TABLE B Time of appearance of pre-natal characteristics in relation to weight and crown–rump length of the foetus of the pony

Days after end of oestrus	Characteristic	Weight (gm)	Approx crown-rump length (mm)
10	Yolk-sac placenta		—
24	Heart beat		5
	Prominent allantois		
	Visible forelimb bud		
	Visible hindlimb bud		
30	Chorioallantoic placenta		10
40	Nostrils and eye buds	1.6	25
	External genitalia		
	Endometrial cups		
	Placental attachment to uterine wall		
50	Appendages developing		35–51
	Eyelids closed		
60 to 80	Uterus carried by placenta		75
80	Endometrial cups start to degenerate		130
100	Endometrial cups completely rejected	82	170–81
160	Hair around eyes and muzzle	1,000	310
220	Hair at tip of tail and mane	3,000	500
300	Completely covered with hair, eyes open	10,000	900

FOETAL MEMBRANES AND FLUIDS

The foetal membranes are the placenta and the amnion. The placenta consists of the chorionic and allantoic membranes which fuse early in embryonic life. The allanto-chorion (placenta) completely surrounds the embryo and contains a fluid (allantoic fluid) which does not come into contact with the embryo itself because of the amnion. This fine, shiny membrane surrounds the foetal foal and remains intact until the final stages of birth. The amnion contains amniotic fluid which bathes the foal but does not irritate or harm it.

The Placenta

The placenta is an organ developed and ideally suited for facilitating exchange between the maternal and foetal blood streams. And because it is not part of the foal's body, but only connected to it by the umbilical cord, it can readily be discarded after birth, when no longer needed. The cavity within the placenta is connected to the foal's bladder through a tube (urachus) in the umbilical cord. Urine formed by the foetal kidneys can thereby escape to the allantoic cavity which acts as an additional bladder able to store the waste products of the kidney until they can be disposed of at birth when the placenta breaks during first-stage labour.

The placenta assumes its role as a means of exchange from about Day 25. Before this time, the developing foetus has been sustained by a temporary organ, the yolk sac. This consists of a pouch containing many blood vessels which grows from the gut of the embryo. The pouch bathes in uterine fluid, extracting the necessary nutrients for the early development of the foetus. By Day 25 the true placenta develops as a very fine membrane filled with fluid. The embryo is at one pole attached by a fine stalk. The yolk sac is becoming redundant and shrinks to an unrecognisable size within the umbilical cord. However, it may remain as a globular amorphous structure, perhaps becoming calcified and resembling a small skull or bony globule. At birth this structure may be mistaken for a miniature twin, but it is really the remains of the yolk sac placenta, vital for the embryo's survival only at the very start of pregnancy.

The true placenta undergoes progressive development, becoming increasingly substantial, until between Days 50 and 60 it develops the fine, velvety surface we recognise when it is expelled at birth. But, before describing the structure of this membrane and its relationship with the uterine wall, we must first consider the endometrial cups, a development in pregnancy unique to the horse family. The cups, as their name suggests, are on the inside (endo-) of the uterus (-metrium). They are saucerlike craters appearing about the 36th day of pregnancy, situated in a ring around the junction of the uterine horn and body, on the side where the foetal sac (blastocyst) has taken its position.

The cups begin as fresh active tissue but by the 90th day they are becoming brown and degenerate. Their active life span is thus about fifty days. For many years the cups have been recognised as a source of a pregnancy hormone PMSG (see page 154). This is present, in large quantities in the mare's blood stream between about Day 40 and Day 110 of pregnancy, forming the basis of the blood pregnancy test.

It was not until the early 1970s that Cambridge scientists led by Dr W. R. (Twink) Allen discovered that the origin of the PMSG hormone was in foetal and not, as previously assumed, maternal cells. The foetal cells burrow into the uterine wall on the 35th day of pregnancy, crossing from a girdle of tissue on the surface of the developing placenta. They remain in the uterine wall until rejected by an immunological reaction similar to that provoked by cells and other matter which is foreign to the body; and, of course, the foetal cells are foreign to the maternal tissues. However, this rejection occurs only after a fairly lengthy period in which the foetal cells remain healthy and actively secreting hormone.

Let us now return to the make-up of the placenta. This is distinct from the equivalent organ in other mammalian species because it covers the whole of the uterine surface from the 100th day onwards. All other

Fig 20 The mature equine placenta, showing the structure of the microcotyledons. Four of these are shown: the two on the left are 'unbuttoned' from their position in the uterine wall (Reproduced by kind permission of D. H. Steven)

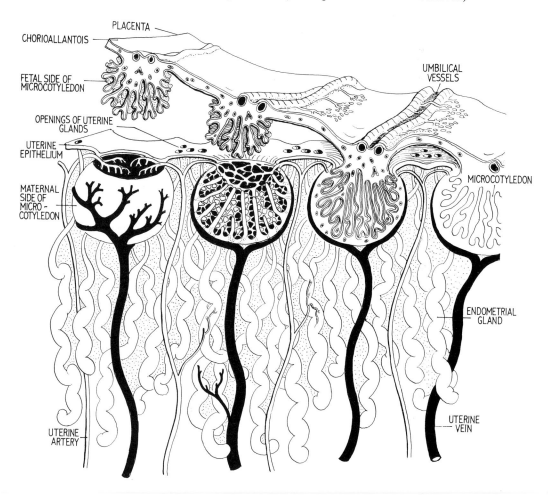

species restrict their placenta to discreet areas of the uterine wall. In woman, the placenta is disc-shaped, in the pig bandlike, and in the sheep and cow, caruncular (buttonlike).

The main practical consequence of this diffuse attachment is that the mare has no spare room for the presence of a second placenta and is not, therefore, naturally able to bear twins (see page 271). Other species have room for two or many more placental membranes. Thus, in the horse, survival in evolutionary terms has depended on an ability to develop not more than one foetus per pregnancy. Twins are a common cause of abortion or undersized and weak foals born at full term.

Another peculiarity of the equine placental membrane is the six cell layers placed between foetal and maternal blood streams. Three of these are placental and three uterine. Thus, if we imagine ourselves travelling from inside a small blood vessel (capillary) in the placenta to a similar vessel on the maternal side, we would pass through (1) the wall of the foetal blood vessel, (2) a layer of supporting (connective) tissue, and (3) the epithelial (outer) layer of the placenta, before we reached maternal tissues. Here we would encounter (4) the epithelial (outer) layer of the uterus, (5) the supporting (connective) tissue layer, and (6) the fine capillary wall of the maternal blood vessel. On entering this vessel we would have passed through the diffusion pathway travelled by the molecules of nourishment and life support, such as oxygen. Those of waste material, such as carbon dioxide, take the opposite route.

In other species, the number of cell layers is reduced, usually by the placenta eroding the maternal layers and, in some cases, losing some of its own. The equine placental membrane, unlike the human placenta, does not erode the uterine tissues; therefore when it peels away after birth it does not cause much, if any, bleeding.

The traditional concept of how the mare's and developing foal's blood vessels came into close contact was that the maternal and foetal sides were linked by fingerlike projections in a simple side-by-side relationship. However, more recent work has shown a much more intricate relationship between foetal and maternal blood vessels (fig 20). The arrangement has something in common with the buttonlike arrangement of the sheep and cow. These can be seen readily with the naked eye but, in the mare, the 'buttons' (microcotyledons) are microscopic. There are many millions of them, covering almost the entire surface of the placental membrane. On the maternal side, ie in the uterine wall, each microcotyledon is composed of an outer capsule with arterial blood vessels entering at the top. These divide into capillaries through which the blood passes into the veins. The veins join together to form one large vein draining blood away at the lower part of the capsule.

On the placental side of the unit, arteries carry blood, driven by the foetal heart beat, into capillaries. These fit into the system of vessels like the clasping of hands. Veins convey the blood back into the foal. Following birth the foetal side of the system is 'unbuttoned' and the placental membrane is expelled from the uterus taking with it no uterine (maternal) tissue.

It has been established that the direction of flow in the maternal and foetal blood vessels is in the opposite direction. This facilitates the exchange of molecules between foetal and maternal capillaries based on the principle of simple diffusion in which gases and substances in solution pass from areas of high to those of low concentration. For example, oxygen in the maternal blood stream is relatively plentiful compared with that in the foetal blood stream, and thus oxygen molecules diffuse from the maternal to the foetal side. Carbon dioxide, on the other hand, accumulates in the foetal tissues and is carried into the placenta, where it diffuses in the reverse direction.

The placenta is also able to exclude selectively some substances or to extract others, according to the needs of the foetus; it can produce hormones and metabolises others; it can form a barrier against many drugs, but unfortunately not all. The placental membrane is a substantial organ in itself, requiring its own nourishment and oxygen.

The Umbilical Cord

The umbilical cord is the link between the developing foal and its essential organ of nourishment. It contains two major arteries and a vein. The arteries are derived from the foetal aorta, the main artery of the heart that carries blood to all the hind parts of the body. However, in the foetus, the aorta possesses two branches that pass in the abdomen to the umbilicus (navel) and thence in the cord to the placental membrane, where they divide into increasingly smaller branches, eventually forming the capillaries of the microcotyledons. The circulation is completed by the veins of the placenta, collecting together to form one large vein in the cord. This umbilical vein enters the foetal body and passes to the liver. Here the blood stream courses through the liver into a further vein that conveys it back to the heart via the large vein (*vena cava*). Thus the foetal heart is able to pump the blood stream, not only through its own body, but through the placenta. The cord also contains the urachus, a duct connecting the foetal bladder to the allantoic cavity.

Allantoic Fluid

This yellowish-brown fluid accumulates in increasing quantities during pregnancy. For example, on the 35th day its volume is only about one-tenth of a pint (60ml), but by Day 60 it is about 2 pints (1,200ml); by full term it has increased to 14 pints (8 litres). The fluid is formed by the placenta and from urine passed through the urachus. It is thus a receptacle for unwanted salts that are not removed through the placental/uterine contact. At birth, the fluid escapes when the placenta ruptures at the 'breaking of the water'. The fluid also contains cells, rubbed off the placenta and amnion, and a somewhat curious structure called the hippomane.

The hippomane is a soft, flat, roughly oval-shaped body measuring, at full term, about $5^1/_2$in x $^1/_2$in (14cm x 1.5cm). It is usually brown or cream-coloured but, in some cases, white. It contains high concentrations of various salts including calcium, phosphorous, sodium, potassium and magnesium. There is usually only one hippomane, but occasionally smaller ones develop. It is probable that the hippomane is formed in a similar manner to crystals. Some cells which have rubbed off the placenta provide a nucleus around which the salts and other substances in the allantoic fluid attach themselves. The term 'hippomane' comes from the Greek word meaning 'horse madness'. It was once thought, and is still

The hippomane (centre) with parts of the afterbirth

believed by some, that the hippomane may be used to influence a stallion's behaviour.

The Amnion

The amnion is the shiny white membrane first seen surrounding the foal at birth. It develops around the 17th day of pregnancy and is complete by Day 21. It consists of a fine transparent membrane interlaced by a network of blood vessels. It forms a protective cloak around the foetus, preventing any contact with the allantoic fluid and also, at birth, acting as a means to reduce friction between the foal and the birth canal.

Amniotic Fluid

Amniotic fluid is a clear, straw-coloured, slightly viscous fluid containing cells rubbed off the foal's skin. Its volume increases from about 1 pint (600ml) in the first third of pregnancy to about eight times this volume at full term. It is formed partly by the blood vessels of the amnion and partly of secretions from the body orifices, ie the mouth and urinary tract. It does not normally contain faeces. Its composition is such that it is ideally suited to bathing the airways of the foetal head and neck, the eyes, ears and skin. It may also be swallowed, thus lubricating the gastro-intestinal tract. It probably plays a dynamic part in the development of the stomach by providing fluid to distend the organ as it develops.

HORMONAL MECHANISMS MAINTAINING PREGNANCY

Progesterone is the essential hormone involved in preparing the uterus to receive the fertilised egg, and for maintaining the union between the placental membrane and the uterine wall. However, evidence suggests that although progesterone is essential, so too are other hormones (eg oestrogen, prostaglandin and prolactin) acting in unison or combination with one another. Hormones control, on the maternal side, (1) the state of the uterine lining, (2) the amount of blood coursing to the uterine surface, (3) activity of the uterine glands, (4) sensitivity of the uterine muscle to contract when stimulated, (5) the degree to which the uterus can expand to accommodate the developing foal while, at the same time, maintaining the cervix in a tightly closed position. On the foetal side, hormones control (1) blood flow in the placenta and (2) the state of the lining of the placenta

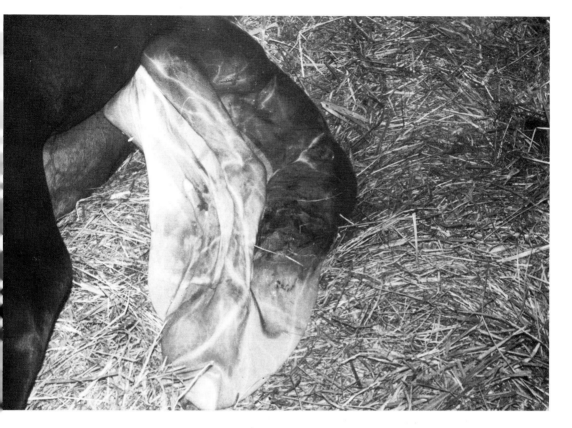

The foal is delivered in its amniotic membrane, providing a slippery surface to facilitate delivery through the birth canal

to match, and form a junction with, its uterine counterpart. Hormones such as oestrogen and cortisol also play a major role in foetal growth and development to full size and maturity.

The way in which hormones act is complicated. Many have opposing actions on specific target tissues. The net effect is, therefore, a balance. In other cases there is a co-operative effect, whereby one hormone provokes or improves the action of the other. Some hormones require another to prime the target cells (ie those on which the hormone is acting) before they, themselves, can have any effect. Others, such as prostaglandins, may have one effect at low and quite another at higher levels – for example, causing small blood vessels to dilate when present in low, but constricting them in high concentrations.

There is a further twist to this complex story. Hormones are substances that are 'built' from other substances and are also broken down into compounds which may be less active, or inert. The build-up process enables the body to form hormones from 'raw material' and from components available in the diet or produced within the tissues. The raw materials may be common to several hormones which in themselves may be broken down and built up into others. For example, oestrogen and progesterone are both built on compounds known as steroids and it is possible, therefore, for the tissues to convert progesterone to oestrogen.

The hormonal system works on the principle that hormones are produced, on the one hand, by glands, and destroyed, on the other, by organs or by the tissues on which the hormones have their action. If there was not such a means of making hormones inactive, they would accumulate in ever-increasing quantities. The sensitivity of the hormonal control mechanisms depends on the ability of the glands to vary their productivity and of the tissues to eliminate the hormones at a rate which is appropriate to the level of production. The reader should appreciate that in practice there is no simple method of approach to diagnosis, preventing abortion or understanding hormonal function and dysfunction.

A further dimension to the story of the hormonal system is that, in the pregnant mare, we are dealing with two distinct beings. Thus there are two separate hormonal systems, each having its own glands (pituitary, adrenals, gonads, etc). In addition, the placental membrane produces hormones and modifies and alters those brought to it from the foetal gonads and elsewhere. This means that the concentrations of the various hormones, such as progesterone, will not necessarily be the same in the foetal and maternal blood streams. We have access only to maternal blood, and measurements made here may not reflect the amount of progesterone or other hormones available to the foetus for the maintenance of pregnancy.

Studies of the hormonal status of the mare and developing foal during pregnancy have developed in recent years along two main lines of investigation.

1 The traditional method of sampling the mare's blood, obtained from her jugular vein, has been extended through the development of sophisticated measuring systems in the laboratory. For many years it has been possible to measure levels of PMSG (see page 143). Nowadays, most of the other sex hormones, progesterone, oestrogen, FSH and LH, can also be measured at extremely small levels, even to a thousandth (milligram)

of a thousandth (nanogram) of a thousandth (picogram) of a gram. Investigators in many countries have reported changing hormonal patterns on a daily basis throughout the whole of pregnancy.

2 Studies on the foetal side of the placenta are technically difficult. Techniques have been developed in sheep to enable the foetal and maternal blood streams to be sampled concurrently. Fine-bore tubes (catheters) are placed in the umbilical and uterine blood vessels during pregnancy, by a surgical operation under general anaesthesia. The other ends of the tubes are brought through the abdominal wall and overlying skin to the exterior. Following recovery from the operation, samples of blood may be withdrawn at regular intervals without disturbing mother or foetus. In this way scientists have direct access to blood on both sides of the placental barrier. The technique has been used for many years in sheep, providing a great deal of useful information about hormones and metabolism of the foetus at differing stages of pregnancy. Unfortunately, the horse has proved a much more difficult subject on which to use this technique. Mares often abort within about twenty to thirty days after the operation, and this has made it possible to conduct studies during only relatively short periods of pregnancy.

The use of radioactive substances injected into the foetus, followed by long-term studies of the appearance of the radioactive labelled material in the mare's excretions, is a technique that has yielded some interesting information. Foetal gonadectomy (removing the ovaries or testes of the foetus) and observing the effects on pregnancy and foetal development has also contributed to our knowledge.

In recent years, the clinical approach to the foetus itself has progressed. Amniocentesis involves taking a sample of the amniotic fluid by means of a needle inserted through the abdominal wall. Allantoic fluid can also be obtained by this means. The technique has been practised for many years in human patients and its recent introduction in animals is based largely on the development of ultrasound. The mare is sedated with a suitable tranquilliser, and local anaesthetic placed in the abdominal wall after the position of the foetus has been identified, together with its surrounding membranes and fluids, by ultrasound scanning (page 350). Preliminary studies have also been undertaken of techniques whereby the foetus itself is injected with various drugs such as ACTH. Withdrawing fluid for analysis and injecting the foetus are, so far, techniques employed for research purposes. However, over the next few years, advances may allow approaches to be used for clinical diagnosis and therapy.

The sum total of these investigations still leaves us with many pieces missing from the jigsaw puzzle of the hormonal control of pregnancy in the mare.

Progesterone

Progesterone receives the distinction of being described as the hormone of pregnancy because it is the one hormone known to be essential for the continued well-being of the foetus. The mare's ovaries produce progesterone up to about Day 120, and the foetus or placenta contributes from the 40th day onwards to full term. The ovaries of the mare may be removed, experimentally, and abortion not occur once the foetal source has been established. Maternal levels of blood progesterone are derived from the yellow body formed at the time of conception. This yellow body would normally have its life ended by a secretion of prostaglandin from the uterus after about fifteen days. However, because of the presence of a fertilised egg, its life is extended for up to one hundred days. We do not, as yet, understand how the mare 'knows' that she is pregnant and by what signal or by what means the yellow body is protected. It is still sensitive to the action of prostaglandin, and the mechanism of its continuing life does not appear to depend on its protection from the action of this hormone. Rather, it seems that there must be some suppression of the release of prostaglandin from the uterus by the presence of the developing foetus. The trigger mechanism (switching-on process) which turns the short-lived into a long-lived yellow body takes place around the 14th day of pregnancy. The results of recent work suggest that two factors are involved in the suppression of prostaglandin secretion in the uterus, thereby prolonging the functional life of the yellow body formed at the time of conception. The first of these mechanisms is the movement of the embryo around the uterus up to Day 16. This is discussed in more detail in chapter 15. It is of comparative interest that, in cattle, the embryo moves around one horn only whereas in the horse it travels around both horns and the body. This difference corresponds to the fact that prostaglandin passes locally from the uterus to the ovary, whereas in the horse it passes into the blood stream. In the cow, the embryo occupies the horn on the side of ovulation but in the horse it may occupy either horn. The contact of the bovine embryo with the one horn therefore blocks the mechanism of prostaglandin release on the side of the uterus where it is important, whereas in the horse the contact has to be over the whole surface of the uterus to achieve a similar blockage. The second mechanism of blockage is thought to be protein secreted by the embryo that establishes recognition of pregnancy by the mare from Day 16. The exact function of this protein is not yet known.

Secondary or accessory yellow bodies form around Day 25, as a result of ovulations occurring under the influence of the pituitary hormones FSH and LH, surges in the levels of which occur even though the mare is pregnant. It was once thought that these secondary ovulations occurred as the result of the high levels of PMSG which begin to appear in the mare's blood stream from about Day 40 onwards. However, recent work has shown both that this ovarian activity occurs before PMSG levels have increased, and that the ovaries do not respond to PMSG (see page 143). The accessory yellow bodies have a variable life but cease to function around the 120th day of pregnancy. Progesterone levels in the mare's blood stream fall and remain low throughout the rest of pregnancy, until they increase in the days before birth. The source of progesterone, after Day 150, is entirely due to the transfer of placental and foetal sources, and not to any activity in the mare's ovaries.

Oestrogens

There are a number of different substances included under the term 'oestrogens'. These include oestradiol, oestrone, and two compounds peculiar to the horse family, equilenin and equilin. These last two substances appear to have no hormonal significance but can be harvested commercially and are used in face creams sold as cosmetics. Oestrogens appear in large quantities in the mare's urine from Day 120 onwards. Levels reach a peak in the middle of pregnancy and then decline towards birth. These oestrogens are 'built' on compounds produced by the foetal gonads (see page 150). Oestrogens are probably necessary for the growth and development of the foetus but not for the maintenance of pregnancy. Measurement of oestrogens in the mare's urine during pregnancy is the basis of the urinary pregnancy test which is accurate from 120 days onwards.

Equine Chorionic Gonadotrophin (eCG)

Pregnant mares' serum gonadotrophin (PMSG) has recently been renamed equine chorionic gonadotrophin (eCG). This has follicle-stimulating and luteinising properties. However, these hormones are produced, not by the mare's pituitary as they are during the oestrous cycle, but by special cells that develop in the endometrial cups. eCG can be obtained from pregnant mares' blood between about Day 40 and Day 100 of pregnancy. It may be purified and used commercially as a drug for its follicle-stimulating and ovulatory properties. The curious fact is that, although this drug is effective in other species, including humans, it has no effect whatsoever on the mare's ovaries. Why then, it may rationally be asked, is it present in such large quantities in pregnant mares? No one knows the exact answer to this, but recent work suggests that the FSH and LH components of eCG are sufficiently different from their pituitary counterparts to be recognised as such by the ovaries of the horse – but not those of other species. It is thought that the large quantities of eCG may have a part to play in suppressing the immune response of the early pregnant uterus so that it does not reject the developing foetus. Further, it may stimulate the development of the foetal gonads and account for their great increase in size during the period that the hormone eCG is being secreted.

eCG is secreted by the endometrial cups between Day 7 and about Day 100 of pregnancy. The cups, so called because of their shape, are areas in the uterine wall where cells migrate from the girdle surrounding the embryo (see page 143) around Day 37. The cups cover a small circular or oval area and a series of them is to be found encircling the base of the uterine horn on the side where the embryo is developing. The mare's pituitary continues to produce FSH and LH even though the mare is pregnant. These pituitary hormones are probably associated with the development and ovulation of follicles forming accessory yellow bodies (see above). It is not known if these surges in FSH and LH levels occur throughout the whole of pregnancy, but they are present at birth and in these circumstances are probably responsible for priming the follicles, present at the foal heat.

Prostaglandins

There are a large number of different prostaglandins (PG) besides those responsible for terminating the life of the yellow body. The actions of the PGF and PGE series range from causing changes in the size of the smaller blood vessels to contracting or relaxing muscle fibres. These substances are, therefore, closely concerned with changes in blood flow and uterine muscular activity and they are present in the foetal circulation and in allantoic fluid. Levels tend to increase towards full term, but marked rises in concentration occur only during delivery of the foetus at birth. Prostaglandins are part of the intricate hormonal mechanism of birth and, also, of placental and foetal function during gestation.

Cortisol

Cortisol levels in the foetus remain low throughout pregnancy and rise only to a small degree towards the time of birth. This hormone plays a significant part in foetal metabolism and in the birth process, together with hormones such as prostaglandin and oxytocin.

PREGNANCY DIAGNOSIS

Pregnancy diagnosis is needed during the breeding season to assist management, and at sale times, for insurance, meeting contracts for 'no foal no fee' nominations, when planning the following year's matings and, most importantly, as a first step to the diagnosis and treatment of infertility.

There are four main methods of examination: ultrasound scanning; rectal palpation; blood test and urine analysis.

Ultrasound Scanning

This technique is discussed in detail in chapter 14.

Rectal Palpation

19–21 days. The rectal examination may be performed as early as the 17th day after the ovulation in the previous heat period, ie 19 days following the last service. At this stage the findings are (a) the presence of a foetal sac in a uterus with turgid horns, (b) no foetal sac felt, but the uterine walls are turgid, (c) the uterine walls have some tone but are not particularly turgid, and (d) the uterine walls are slack and with no tone. These findings are reported as (a) definitely in foal, (b) probably in foal, (c) probably not in foal, and (d) definitely not in foal.

It is important to appreciate that this very early diagnosis cannot be 100 per cent definite, but provides very useful information to stud managers during the breeding season. A knowledge of the probable sexual state of the mare at this vital occasion can be very helpful and may enable stud management to take special precautions in cases where a mare is suspected of 'cheating', ie not in foal but also not showing at the time of the expected next heat following mating.

30–35 days. A pregnancy examination performed 30 to 35 days after service is more definite because the enlarging foetal sac is more readily palpable. The advantage of an examination at this stage is to determine if twins are present, so that they can be removed before the endometrial cups become active. The presence of PMSG usually suppresses normal cyclic activity.

39–42 days. At 40 days the foetal sac is sufficiently large for an accurate diagnosis of pregnancy to be made in all cases. In some instances the sac may feel small or of less tone than normal. The vet reports this finding and probably suggests a re-examination in a week or ten days to determine if the pregnancy has continued normally.

60 days. During the breeding season a re-check at 60 days is advised as a small proportion of individuals lose their pregnancy between 40 to 60 days. In the case of twins, the sacs are by now lodged against each other in the uterus, and it is difficult to distinguish one from the other. By 100 days onwards, it is impossible to diagnose twins by the manual examination. Some people fear that the manual test may precipitate abortion. There is no evidence that abortion follows an examination performed at any stage of pregnancy. The technique is widely employed, and the very great majority of mares examined continue their pregnancy and deliver a live, healthy foal at full term. In practice, attempts to dislodge one of twin foetal sacs are frequently unsuccessful and both continue to develop despite the pressures brought to bear on them: a very much greater force being exerted than would normally be the case in a routine pregnancy diagnosis.

However, it would be wrong to state categorically that handling the uterus, or even restraint of the mare for the examination, *cannot* cause abortion in particular individuals susceptible to the effects of these procedures. There are alternative techniques that may be employed if it is supposed that the manual test might be harmful in a particular case or if the owner of a mare has reservations about the method.

Blood Test

45–90 days. The blood test is based on the detection of PMSG and is accurate from the 45th to the 90th day of pregnancy. Positive results are accurate outside these dates, eg 40 to 45 and 90 to 120 days, but negative results during these periods may be false. The test was developed as a biological method using mice, rabbits or frogs, but is now mostly conducted in test tubes using immunological techniques. The test is extremely accurate. However, it is possible in a small number of cases to have a positive result but no subsequent foal. The reason for this discrepancy, affecting about one in ten pregnancies, is that the cells that secrete

PMSG are 'turned on' between Days 35 and 40 and continue to secrete irrespective of the presence or absence of the foetus, until they are finally rejected at about 90 to 120 days. The ten per cent discrepancy is sometimes assumed to be an inaccuracy in the test but, in fact, the test for the hormone is 99.9 per cent accurate, and it is only the biological quirk of nature which makes it seemingly unreliable.

Urine Test

The urine pregnancy test is accurate from 120 days of pregnancy to full term. It is based on the chemical detection of the hormone oestrogen, which appears in the urine during this period. The test is highly reliable but there are instances where equivocal results may be obtained and a re-test may be necessary to confirm the result in these cases. The disadvantage of the test is, of course, the problem of collecting urine from unco-operative subjects.

Other Methods

There are other methods, not usually employed in practice, to diagnose pregnancy. Electrocardiography (ECG) may be used in the second half of pregnancy to record the foetal heart beat. The electrode plates are placed on the mare's back and abdomen, and the foetal beat may be identified on the tracing of the maternal heart beat. Heart beats may also be detected by means of an ultrasound probe which can also be used to monitor foetal heart rate and movement. The probe is placed in the mare's rectum and records any change in position of the foetal appendages and body. Radiography has also been used to determine foetal movements in pony mares using high-powered X-ray equipment. It is doubtful if any of these methods are of particular value in practice but they have research potential for studying the health of the foetus and the relationship between foetal well-being and changes in the maternal environment. Unfortunately, they cannot be used to determine the presence of twins, although if two foetal heart beats are recorded this is definite evidence of twins. However, the second heart beat may not be visible or may be mistaken for a single beat.

PREPARATION FOR BIRTH

Pregnancy, as we noted at the beginning of the chapter, is a period of development and preparation for life in the outside world. The whole period of foetal development is built, brick by brick, towards a completed structure of maturity. A foal delivered without hair would, obviously, be fatally ill-prepared to survive even in a summer climate. Hair grows to form a coat well before foaling, and other organs, such as the lungs and nervous system, are necessarily fully developed by foaling time. Most, although not all, the organs are functioning by birth. The heart and circulatory system, as we have seen, are performing the same work before as after birth, albeit at a lower pace. The movement of the foetus, including movements of the chest stimulating breathing, plays an important role in conditioning the muscles and nervous system to the work they will have to undertake as soon as the foal is born. All systems, including those of metabolic liver enzymes and glands supporting hormonal and digestive activities, must be primed at birth, ready to undertake the task of meeting the challenges of the world outside the uterus. Inside the uterus, the foal is protected by its dam; outside, the body must fend for itself. It is only in the last stages of pregnancy that all the systems are fully developed. To be born with any one of them deficient (eg minus a liver enzyme or with insufficient lung capacity) would present the newborn foal with a problem akin to towing a car uphill with a chain in which there was one weak or missing link. Such a foal we would call premature or immature.

To be regarded as normal, pregnancy must last for more than 320 days. A foal born before this date will be premature, ie underdeveloped in size and vital body systems. Foals born between 320 and 355 days are said to be full term, ie in a period of maturity. Beyond 355 days the pregnancy may be considered as prolonged. Some abnormality of growth may have occurred and, although the produce may be mature, the survival potential of the foal is reduced. This may also apply to full-term foals where the placenta has been damaged and development has been impaired in consequence. Some of these individuals show signs of immaturity and are referred to as being dys- (bad) mature. Human infants in this category are described as being 'small for dates'. We will return to the subject when discussing diseases of the newborn foal (see page 297).

The relevance of the discussion here is that foaling takes place normally only when the foetus is mature and ready to leave the uterine environment. In mares this is between 320 and 355 days from last service by the stallion. Normal gestational lengths vary considerably, and an individual which requires 355 days for development would undoubtedly be immature if delivered at, say, 325 days. How, it may be asked, do the mare and foetus know when foaling should take place? The timing of birth depends, to a

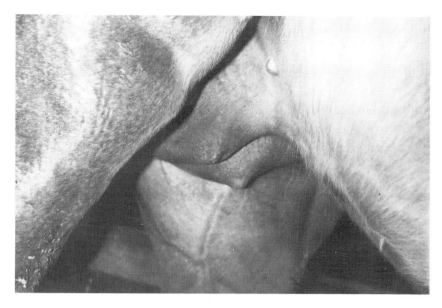

(above) *The state of the mammary glands approximately 1 week prior to foaling;* (below) *a fully-developed udder photographed while the mare is delivering her foal*

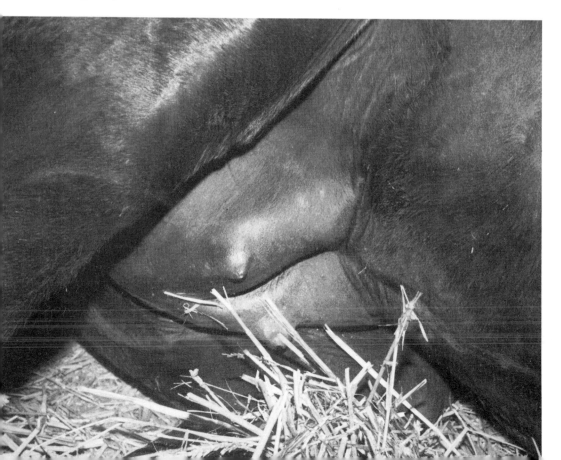

large extent, on the foetus, which appears to have the ability to signal to the mare that it is mature. We shall learn, in the next chapter, that hormones play a dominant role in determining the process of foaling; but the mare is able to control the event to the extent that most foalings take place at night. Thus, it appears, the foetus controls the overall length of pregnancy, and the mare has the power of fine tuning, to the hour of night when she delivers. In this way, maybe, nature has provided the ability to deliver in darkness when predators are asleep.

In the weeks prior to foaling the mare's mammary glands (udder) start to enlarge as the glandular substance increases. Secretions of milk (known as colostrum or first milk) collect in the glands. They contain a high level of protein and, in particular, the protein known as globulin composing the protective substances or antibodies essential to the well-being of the newborn foal.

6

FOALING

Birth is a unique event of mammalian life, whereby the foetus is expelled from the security of the maternal uterus into the harsh realities of a world outside. The change in lifestyle, for the foetus, is dramatic; and it inevitably presents a challenge to survival which we will consider in the next chapter. Here we are concerned with the birth process itself.

It is convenient to describe birth by three stages relevant to biological events that form a progressive sequence. First-stage labour is the preparatory contracting of the uterus, second-stage marks the actual delivery of the foal through the birth canal, and third-stage the expulsion of the membranes (afterbirth), no longer required for foetal life.

Hormonal Control of Foaling

Foaling, as we saw in the last chapter, is a predetermined event based on the maturity of the foetus and the ability of the mare to select the hours of darkness for her 'confinement'. Hormones initiate and control the process of birth, acting on the muscles of the uterus to secure contractions that are the basis of the forces which deliver the foetus through the birth canal. Hormones are also responsible for changes in the genital tract, such as the relaxation of the cervix, making it possible for the foetus to pass through the canal to the outside.

Once the contracting uterine muscles cause the foetus to start its journey to the outside world, delivery is supplemented by powerful straining of the mare's abdominal muscles, as she lies on her side. The newly activated uterine walls and lining of the birth canal 'dispatch' nervous impulses (messages) to the brain and spinal cord which cause the reflex behaviour that we associate with foaling (fig 21).

Two hormones, oxytocin from the pituitary and prostaglandin from the uterus and the placenta, are responsible for the powerful contractions of the uterine muscle already made sensitive to their actions by the presence of the hormones oestrogen and cortisone. Another influence contributing to the contractions of uterine muscles is the reducing levels of progesterone at this time. Progesterone has the opposite effect to oestrogen by making the uterine muscle less susceptible to the action of oxytocin. The

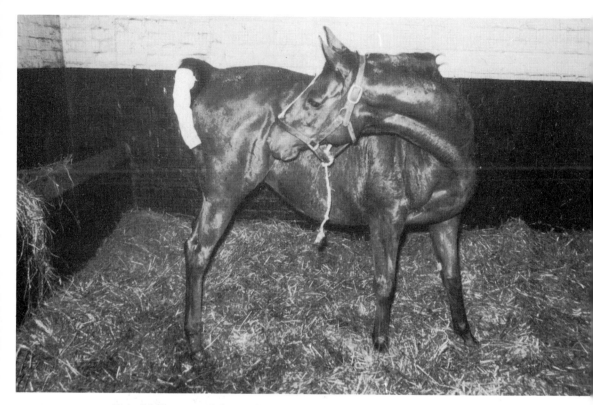

Pain felt because of the uterus contracting causing signs of first-stage

Second-stage is the period in which the foal is delivered

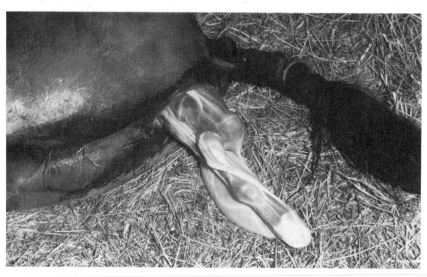

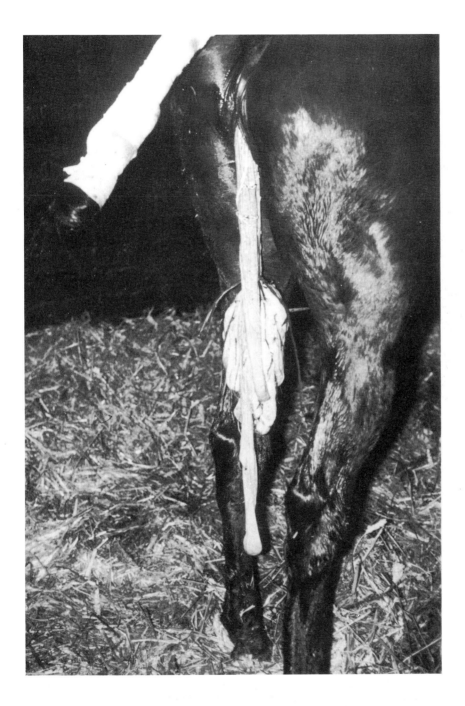

Third-stage marks the expulsion of the afterbirth

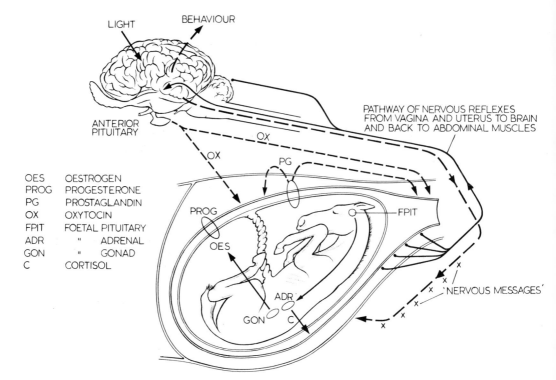

Fig 21 Birth is controlled by hormones from the mare's and the foetal glands acting on the uterus, cervix and vagina of the mare. Reflex nervous activity on the part of the mare is involved. The foetal pituitary (FPIT), adrenals (ADR) and gonads (GON) combine to act along pathways illustrated, with the mare secreting hormones on the uterine side of the combination of maternal/foetal control of birth

hormonal changes and interactions are too complex to be discussed here, and there are still many gaps in our knowledge.

THE ANATOMY OF BIRTH

The Birth Canal

Birth involves the separation of two beings, the one passing out of the other. We are concerned here with the structure and arrangement of the two individuals, the mare and the foetal foal, during this process of separation. The foal has been lying in the uterus on its back with its head and legs flexed (fig 22a); it now has to pass from the uterus to the outside through the mare's pelvic outlet (fig 22d). The passage of the foal to the outside takes it through the cervix, vagina and vulva, soft structures which can expand sufficiently to accommodate its presence. The birth canal consists, therefore, of a soft tube surrounded by an unyielding bony hoop (figs 23, 24). The hoop, being unyielding, determines the ease of passage of the foal during birth. The opening provided by the hoop is roughly conical in shape. It is formed by the sacrum and the first three tail vertebrae above the shaft and floor of the pelvis on either side and below. The opening at the uterine end is somewhat larger than the outlet. In a Thoroughbred mare the vertical diameter of the inlet is about 9.5in (24cm) and of the outlet about 7.9in (20cm). The largest diameter is slightly

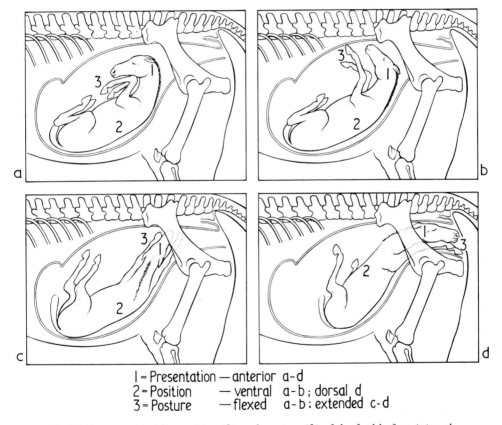

1 = Presentation — anterior a-d
2 = Position — ventral a-b; dorsal d
3 = Posture — flexed a-b: extended c-d

Fig 22 Presentation (1), position (2) and posture (3) of the foal before (a) and during first-stage labour (b and c) and in the early part of second-stage (d) (From Jeffcot and Rossdale – Supplement 27 of the Journal of Reproduction and Fertility)

oblique to the mid-line. This arrangement is important relative to the position of the foal as it enters and passes through the birth canal.

During first-stage delivery the cervix dilates, the vulva relaxes and the vaginal secretions become moist. The soft structures of the canal do not obstruct the foal's journey through the birth canal but, in fact, contribute to the ease of its passage across their surfaces. The bony structures, however, by their very nature cannot assist, except to the small extent that the ligaments binding them relax at the time of birth and thereby slightly enlarge the hoop they form. It is, therefore, necessary for the foetal foal to align itself in such a way as to reduce its bulk and to streamline its appendages to minimise the risks of becoming stuck against the bony pelvis. The way in which the foal achieves this most advantageous alignment is illustrated in fig 22.

Foetal Alignment

It is customary to define the alignment of the foetus in terms of *presentation, position* and *posture* (fig 22). *Presentation* refers to the region of the foetal body presented to the pelvic inlet. Normal presentation is for the foal to come head first. *Position* refers to the alignment of the foal's body relative to the mare's vertebral column. Thus, the correct position is for the foal's back bone to be uppermost and applied to that of the mare. 'Upside-down' means that the foal's chest is uppermost. *Posture* denotes the disposition of the head, neck and extremities relative to the foetal body. In foetal life these are, as already mentioned, flexed, but during delivery they are extended.

The foetal foal, in the sixth month of pregnancy, comes to lie in an anterior presentation and ventral position (ie with its head towards the cervix and lying on its back). Its head, forelimbs and hind limbs are flexed and it maintains this alignment until birth. It has been shown by radiography that the foetus makes apparently 'purposeful' movements during first-stage labour, just before the start of delivery. It extends its head and forelimbs and turns from the ventral 'upside-down' position to the dorsal 'correct' position. These 'purposeful' movements are probably the result of foetal reflexes initiated by the contracting maternal uterus. However, other influences include changes in the mare's posture as she gets up and down, lies on one side and then on the other, and by pressure changes as the placenta ruptures through the opening cervix. These pressure changes not only help to re-align the foetus but also to push it towards the inlet of the birth canal. We cannot precisely determine the

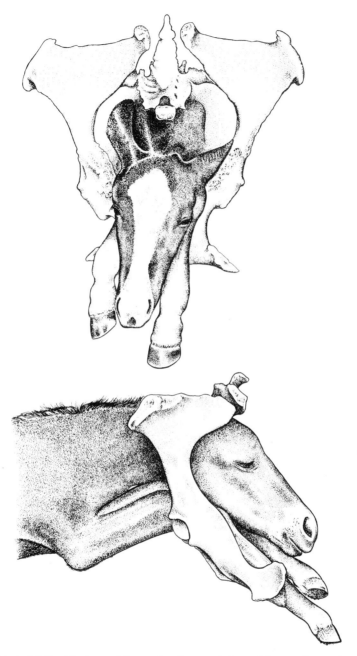

Figs 23 & 24 The birth canal is a soft tube consisting of the cervix, vagina and vulva, surrounded by the bony hoop of the mare's pelvis through which the foal must pass during second-stage labour

extent to which gravity, foetal reflex activity and changing pressures contribute to the process; we do know that the foetus turns during first-stage and extends its limbs to bring itself into the optimum alignment for passage through the birth canal during second-stage (fig 22). The movement of the foetal muzzle and forefeet pressing against the cervix probably plays some part in causing this structure to dilate, during late first-stage labour.

The Forces of Delivery

The wall of the uterus contains muscle which contracts rhythmically in waves which work backwards but advance from one end of the uterus towards the cervix. During first-stage, the waves passing down the uterine horns and body towards the cervix exert increasing pressure on the contents of the uterus, ie the foetus and the fluids surrounding it, thereby ensuring that they are squeezed towards the cervix. These muscular contractions are painful, and are responsible for the signs shown by the mare. These signs, as we shall see later, include sweating and uneasiness. The uterine contractions increase in severity and effect. The cervix dilates under the mounting pressure in the uterus and the presence of the foetal appendages pressing against it. The increased pressure and the opening cervix cause the placenta to bulge into the cervical aperture. The membrane is particularly thin at its cervical pole and it ruptures naturally once the process of first-stage has shifted the foetus into the entry of the birth canal. Allantoic fluid escapes and this, the breaking of the water, marks the onset of second-stage, ie delivery. Second-stage delivery is characterised by contractions of the abdominal muscles, ie so-called voluntary straining, in which the mare lies down, takes a deep breath, thus holding the rib cage and diaphragm at the position of maximal inspiration, and uses her abdominal muscles to bear down on the abdominal contents including the uterus (fig 25). The pressure exerted by this straining is considerable, rising to many hundreds of millimetres of mercury (compare ordinary blood pressure of 120mm Hg). The only escape for this pressure is through the weakest point of the abdominal cavity, which at birth is the opening of the cervix. All the pressures on the abdominal contents are, during straining, directed towards this one exit. The pressure will, of course, empty the rectum in the manner of defecation, which is the reason why mares pass faeces during second-stage. Clearing the rectum prevents faeces from obstructing the upper part of the birth canal. The action of the mare in lying down minimises the effects of gravity which,

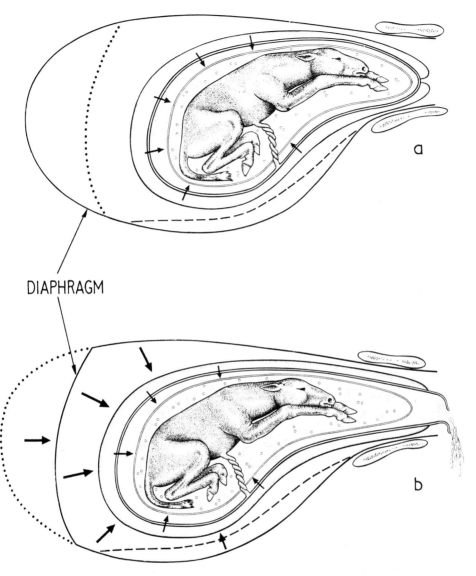

DIAPHRAGM

Fig 25 The forces brought to bear on the foetus during first-stage (a) through the waves of muscular contractions passing through the uterine wall. In second-stage (b), the cervix dilates and the placenta ruptures, allowing the escape of allantoic fluid. The uterine forces are supplemented by contractions of the abdominal muscles. These forces are distributed throughout the abdominal cavity and their effect is increased by the fixing of the diaphragm in a position of inspiration (breathing in). The effect of delivery is thus one of pushing the foetus through the birth canal

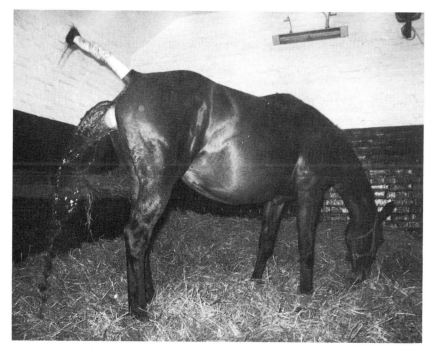

The moment of 'breaking water' and the start of second-stage

if the mare were straining and standing, would pull the foetus back into the abdomen against the direction of straining.

In Summary

The combined force of the uterine and abdominal muscles act to *push* the foetus through the birth canal, lubricated as it is by allantoic and amniotic fluids. It is important for the reader to recognise that the natural forces of birth are those of pushing and not of pulling the foal through the birth canal. Any assistance we can provide the foal during delivery is necessarily one of pulling, and on these occasions we should remember that such assistance must be regarded as an unnatural and not a natural force. It cannot be too highly emphasised that strong traction on the forelegs can damage the foal's chest during delivery.

THE THREE STAGES OF LABOUR

First-stage Signs

First-stage contractions cause the mare to show signs related to the pain felt. These include patchy or profuse sweating, uneasiness, pawing the ground, looking round at the flanks and showing the Flehmen's posture of retracting the upper lip. Some individuals may back against a partitioning wall or cringe in response to the stabbing sharpness of the pain. Sometimes a mare may repeatedly get up and down during first-stage, but this behaviour is usually confined to second-stage. The signs vary in intensity, possibly as a result of the response of the individual to the presence of pain and partly to the actual intensity of the pain. The signs may be very obvious in some and not in other individuals. In practice, some mares hardly turn a hair before they start second-stage. Others work themselves into a frenzy of lather and restless activity, walking or even running round the box in extreme anxiety. Similarly, there are individual variations in the length of second-stage, from a few minutes to many hours. Signs of first-stage appear in waves corresponding with the muscular activity of the uterus; and increase in vigour towards the start of second-stage. However, although we may expect mares in first-stage to show corresponding intensity in their response to pain through the course of first-stage, in practice some mares become uneasy and sweat, followed by a short period of remission during which they may eat or stand placidly, until the next contractions occur.

Some individuals show obvious signs of first-stage and then cool off for hours before, apparently, starting the process all over again. These false alarms may occur frequently before foaling actually takes place. Thus, some individuals may show signs of first-stage on several occasions, hours or days prior to foaling. This is not abnormal and it does not seem to harm the unborn foal.

Prolonged first-stage signs may indicate that the placental membrane is abnormally thickened at the pole, that is the part adjacent to the cervix. In these cases the mare may lie down and strain before she breaks water. The reason is that the abnormal thickness of the placenta prevents it rupturing in the natural manner. The membrane, in these cases, may appear at the lips of the vulva prior to the start of second-stage. If this happens, the membrane should be ruptured by hand or with a pair of scissors. The mare's behaviour of getting up and down repeatedly may indicate that the thickened placenta is causing problems or that the foetus is lying in the wrong position or posture. This can only be determined once the placenta has ruptured.

Second-stage Signs

Second-stage starts at the rupture of the placental membrane and the escape of allantoic fluid (breaking of the water). This may occur when the mare is standing or as she lies down. The allantoic fluid escapes from the vagina as a small trickle, or copious stream, of yellow, brown or yellow-brown liquid. The event is usually recognised with ease by those present, but some mares may start second-stage with so little fluid escaping that the start of this phase is missed.

Once the placenta has ruptured there is no retreat; and the mare must deliver her foal within, at the most, an hour. Most mares complete second-stage delivery within twenty minutes of onset, delivery being accomplished more quickly in older than in younger mares, especially those foaling for the first time.

The amnion should appear between the vulval lips within about five minutes of the placental membrane breaking. The mare usually lies down and may start visible straining at this time. Expulsive efforts in the standing position are usually the result of a nervous mare being disturbed by the presence of attendants. About ninety-five per cent of mares make their final expulsive efforts while lying down and spend the majority of the period of delivery in the recumbent position. However, they may change their position frequently, and it seems that this may in some cases be a positive effort on the part of the mare to shift the way in which their foal is lying. Mares that get up and down repeatedly may be experiencing difficulty because their foal is wrongly positioned, and the person responsible for supervising the foaling should check the position and posture of the foal early in second-stage.

A mare's 'exaggerated' activity is not the only sign of abnormal foaling; marked inactivity during first-stage may be due to weakness of uterine contractions, the malalignment of the foetus, or a particularly small amount of allantoic fluid. Some individuals may fail to proceed with second-stage without indicating that first-stage has actually occurred. For example, a mare may show mild signs of first-stage, lie down and do no more than look round occasionally at her flanks or give a half-hearted strain. In these circumstances the onlooker should check that the placenta has ruptured by inserting a hand into the vagina. We must return to this procedure later when discussing the management of the foaling mare.

The sequence of delivery is illustrated from pages 173–180. It should be noted that one forefoot may be slightly in advance of the other. This is normal and probably helpful in that the relatively bulky elbows do not pass through the pelvic hoop together. Once the foetal head has been delivered

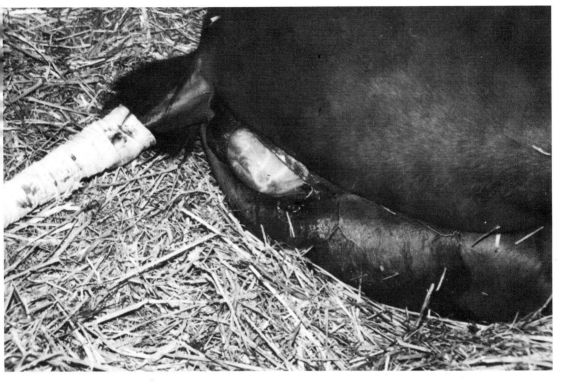

The appearance of the amnion shortly after second-stage has started at the breaking of the water

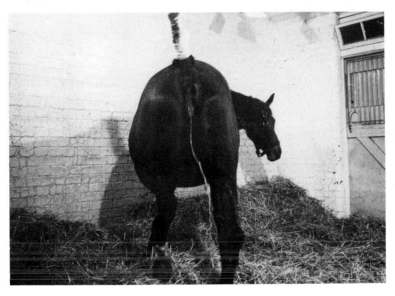

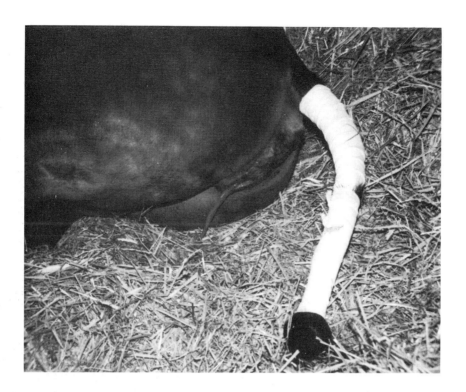

Three examples of breaking water: (opposite, below) *in a fine white stream;* (above) *in a small trickle, together with slimy mucus and the mare in the lying-down position;* (below) *where a light-brown stream is evident. Note the attendant observing the event from the 'sitting-up room' and the mare's tail raised in response to the sharp pain of early second-stage labour*

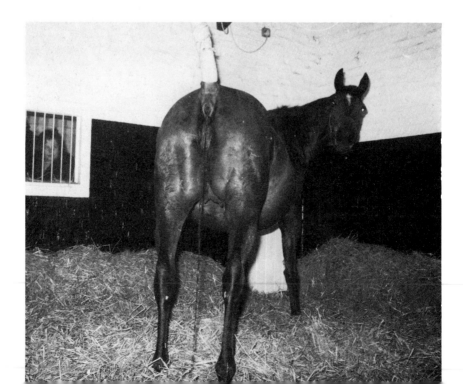

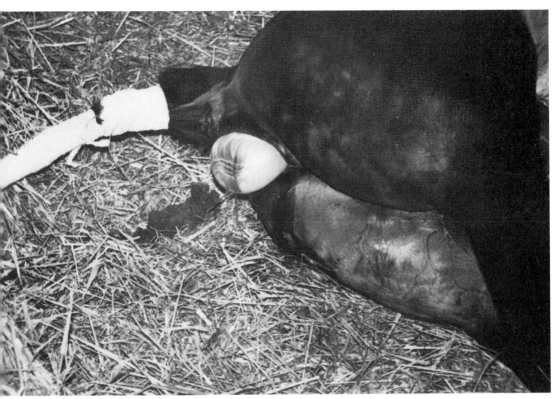

The amnion should normally appear within about five minutes of the onset of second-stage; faeces present in the rectum are expelled by the developing forces of delivery

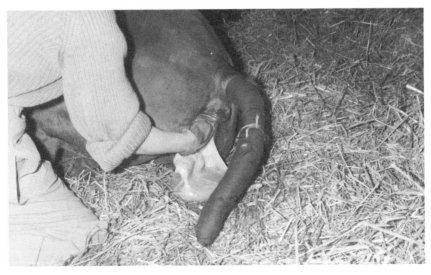

The position and posture of the foal should be checked early in second-stage

A mare in early second-stage labour is showing more inclination to look for the foal, not yet delivered, than to persevere with her labour

(opposite and overleaf) *The sequence of second-stage delivery from the appearance of a forefoot to the complete delivery of a foal. Note the fact that one foot comes in advance of the other and the head is lying above the knees; the foal goes through an arc in that the direction is parallel to the mare's backbone and then arches downwards towards the mare's hocks*

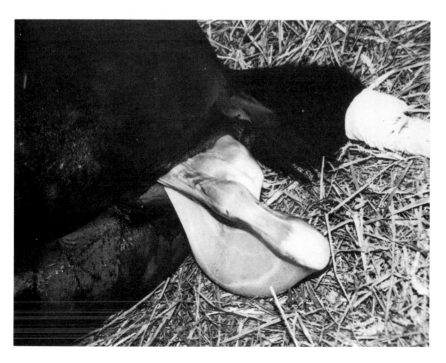

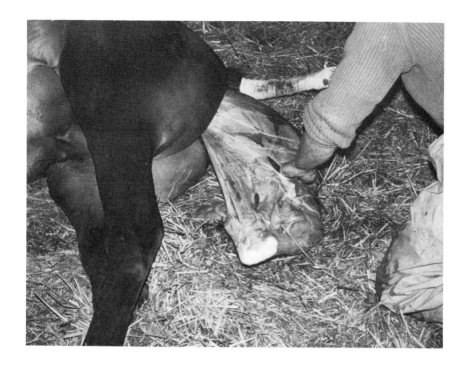

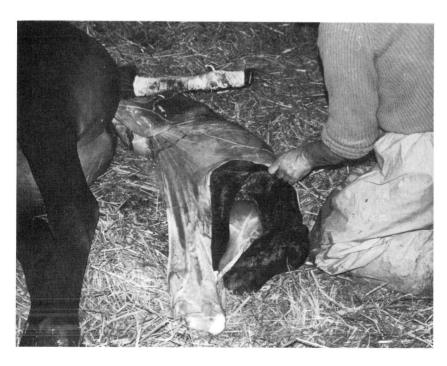

Final stages of delivery take the direction of an arc and this is the direction in which any assistance must be given should it be necessary

past the level of the vulva, delivery is usually completed lying down. If the mare gets to her feet, it is generally because of someone entering the loose box and approaching her. Nervous mares may complete delivery while in a standing position but this is quite the exception and regarded as abnormal. In this case the foetal body is pulled by gravity against the direction of straining. The situation is reversed once the foetal body has passed over the fulcrum formed by the pelvic brim. In this case care must be taken that gravity does not pull the foal too precipitately to the ground.

Some mares find it helpful to press their spine against a partitioning wall during their main expulsive effort. They may roll from side to side or get up and down repeatedly, as if attempting to shift the alignment of the foetus. However, excessive rolling may indicate that abnormal pain is present, for example that associated with internal haemorrhaging (see page 293). Older mares, particularly, may be inclined not to strain even though everything is otherwise normal. These individuals appear to be suffering from a bout of laziness rather than anything more serious – as if waiting for someone to pull the foal from them! But this behaviour may be due to a torn uterus or ruptured gut, or the foetus may be lying in such a way that it cannot be delivered normally. The direction taken by the foetus during delivery lies in the form of an arc as it passes through the pelvic hoop and comes finally to rest at the mare's hocks where the foetal hips become disengaged from the maternal pelvis. At the completion of delivery the foal is lying on its side with its hind legs in the mare's vagina to the level of its hocks. This marks the end of second-stage labour.

Third-stage

Third-stage is the expulsion of the afterbirth. Within about thirty minutes of the foal's final delivery the cord breaks as the mare gets to her feet or as the foal pulls it while struggling. However, the placental membrane remains attached to the uterine wall, gradually separating until it becomes completely free, usually about one hour after the end of second-stage; sometimes it takes longer (up to ten hours) for the membrane to separate from both uterine horns. After ten hours, a retained placenta requires veterinary treatment.

The placenta normally invaginates as it peels away from the uterine wall. It therefore leaves the mare with the smooth allantoic surface outermost, although in some cases the placenta separates from behind forwards, ie from the body to the horns, and leaves the mare with the red, velvety surface outermost. This occurs more often in abnormal than in normal births.

The afterbirth in a Thoroughbred at full term weighs about 20lb (9kg). The placental membrane may weigh 9lb–14lb (4kg–6kg) and its surface area is in the region of 1,015sq in (13,000sq cm).

Third-stage labour is completed when the afterbirth has finally come away from the mare. The mare may show some evidence of pain during this stage by rolling, sweating or pawing the ground, but these bouts of pain are not usually prolonged or severe. They must be distinguished from complications of haemorrhage, and if they persist veterinary advice should be sought.

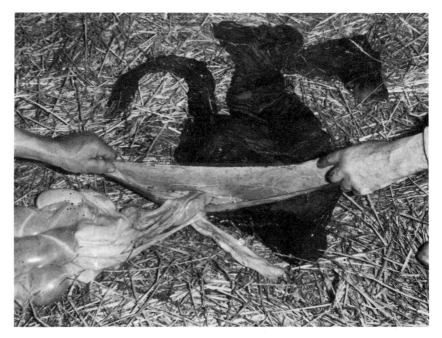

A placenta showing the surface which was attached to the uterine wall. The horns are seen above, and the cord emerging from the opening through which the foal passed is seen below. This opening represents the part that is ruptured at the cervical pole and is held by the hands

MANAGEMENT OF FOALING

We have so far followed the foaling mare through the eyes of the physiologist and obstetrician. It is essential for all those responsible for the care and management of the mare at foaling, vets and non-vets, to understand the basic natural patterns of foaling. For example, we must appreciate the problems faced by the foetus in negotiating the birth canal before we can assist the foal by pulling on its forelegs in an appropriate manner and direction.

Observation

The key to successful management at foaling is to do the least that is required; and to place the most emphasis on observation to assess whether the particular foaling is normal or otherwise. If it is abnormal we must interfere decisively to set matters aright and, if necessary, call for veterinary assistance. The judgement of when to act and when to stay out of the foaling box depends on the experience of the observer.

Good observation of events is essential for anyone wishing to take on responsibility for foaling. People spend many, many hours on studfarms waiting for mares to foal. The first priority is, therefore, to be able to recognise the signs of first-stage. Even before this, a studman must have some appreciation of whether or not a particular individual is likely to foal in the near future. The reader, to achieve efficiency, must be prepared to study a large number of normal foalings before being able to distinguish the abnormal, and must acquire the capacity to observe that something is wrong at the earliest opportunity. The powers of observation are heightened by attention to detail.

Mammary Development

The mammary gland should be inspected each evening. Although we may confine this inspection to the days prior to the expected time of foaling, eg from Day 320 of pregnancy onwards, it is good practice to look 'under' the mare from 200 days onwards for signs of impending abortion or an unexpectedly early foaling. Stud personnel should be encouraged to make these observations as a routine when leading pregnant mares to or from the paddocks. It is important to note any change in size of the udder and whether there is wax on the end of, or milk dripping from, the teats. Changes in udder size near to foaling must be judged at the same time of day. The mammary glands are always larger in the morning, after the mare has been confined to a loose box, than they are in the evening, following exercise in the paddocks. We must, therefore, in noting a change, compare like with like. An increase in mammary size does not necessarily enable us to predict the imminence of foaling. There are many variations between individuals, and mares foaling for the first time may be expected to develop their mammary size over a longer period (three to four weeks) than a mare that has had several foals (one to two weeks). Some individuals experience mammary development very rapidly before foaling and may give a false impression, even a day or two before the actual event. Oedema (soft swelling) around the udder extending forwards to the chest may be encountered in some mares, especially those foaling for the first time. It is not abnormal but a sign of intense mammary development and, probably, high nutritional intake and reduced exercise. The oedema usually clears after foaling without being a problem. Wax on the teats and 'running milk' are features which denote the imminence of foaling, but do not necessarily indicate that foaling will occur on the night of its appearance.

In the days prior to the expected event, mares grazing in the paddock should be closely observed during the daytime, and if an individual begins to show any characteristic signs of uneasiness or of segregating herself from her companions she should, of course, be marked out to receive special attention.

Mammary Milk Content and Foetal Maturity

Recently, it has been demonstrated that certain indicators of foetal maturity may be identified in the mare before foaling. The first of these is the rise that occurs in progesterone metabolites in her blood stream. This increase is the result of a change in the metabolism of progesterone taking place in the placenta. Progesterone is the dominant substance but is formed from other substances (cholesterol and pregnenolone) under the influence of enzymes; and, once formed, it is changed by other enzymes into substances referred to as metabolites. These metobolites include a whole range of minor substances called pregnanes. These may have some progesterone-like activity but are less powerful than progesterone itself. Nevertheless, they are present in enormous concentrations in the mare's blood stream especially during the last twenty days of pregnancy. Their presence, therefore, may be used as a signal that foaling is due to take place in days rather than weeks. If monitored daily, the results may be used to forecast foaling within a matter of two or three days.

The other signal that is now commonly employed on studfarms is the calcium, sodium and potassium concentrations in milk. Sodium levels fall and potassium and calcium levels rise within hours of full-term foaling. Calcium especially may be used to forecast the imminence of foaling because this is usually the last chain occurring in the milk. A dipstick method for estimating these changes has been developed.

There appears to be some relationship between the rise in progesterone levels, changes in milk content and the readiness of the foetal foal for birth. This relationship has not yet been explained fully but the foetus probably influences and controls many of these biological alterations as it reaches maturity. When these relationships are damaged or interrupted, the result is a newborn foal that is unprepared for delivery and life outside the uterus, conditions we describe as stillbirth, prematurity, immaturity (dysmaturity), etc.

'Sitting up'

In the foaling box, vigilance at night is obviously important because this is the period when foaling is most likely to occur. The diligence of people 'sitting up' is something of which the breeding industry can justly be proud. There are not many walks of life in which this degree of dedication is expected or found. It is remarkable how few foalings go unobserved in those breeds, such as the Thoroughbred, where stud management sets itself the task that every foaling should be observed and attended by qualified assistance.

In most cases the signs of first-stage are clearly distinguishable, but a quite substantial number of mares may foal without any apparent warning. It is these cases which may be missed by those 'sitting up'. In addition to the signs of first-stage the nightwatchmen may find it helpful to feel the mare's coat for evidence of warmth and slight dampness and to observe the lengthening of the vulva and the start of 'running milk'. If first-stage signs are suspected, it is usual for the studgroom or stud manager to be informed.

'Breaking Water'

At this time, the most important event to note is the 'breaking of the water'. The time at which this occurs should be recorded, as it may be helpful information should second-stage be abnormal. The behaviour and attitude of the mare during second-stage may help to indicate that something is amiss. Lying without straining, too vigorous straining, getting up and down repeatedly are indications already mentioned. The shiny membrane (amnion) should appear at the vulval lips within about five minutes of the onset of second-stage, especially when the mare lies down, as she generally does following the rupture of the placental membrane. If the amnion does not appear or if the mare lies without straining, this may denote that the foal is not entering the birth canal, perhaps as a result of a postural abnormality, such as the head turned back or the forelegs remaining flexed (see page 292).

Progress in Delivery

Soon after the breaking of the water it is, therefore, good practice for the attendant to feel in the vagina for the nose and two front feet of the foal. (The hands and arms should be well washed with soap and water before introducing them to the birth canal.) The experienced studman can assess

The position, presentation and posture of the foal may be examined with the mare in the lying down (above) *or standard position* (below). *If any correction is to be made it should always be attempted with the mare in the standing position because the foal is then pulled back into the abdomen away from the pelvis by the forces of gravity, thus providing more space in which to work*

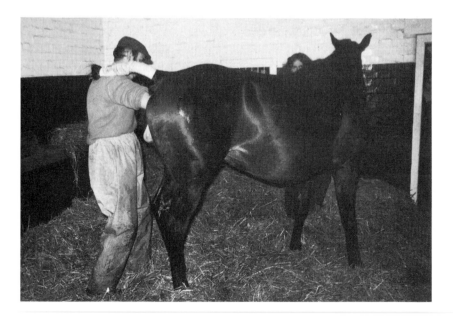

the progress of second-stage and has an intuitive sense of whether or not it is proceeding smoothly and normally. It is not possible to describe, for the uninitiated, the exact determinants of normal or abnormal progress. Individual mares vary in the magnitude of their efforts to expel the foal, according to its size and the ease with which it can pass through the birth canal. Mares foaling for the first time may make the most powerful efforts, but their birth canal is less readily distendable than that of older mares. Judgement depends on the experience of observing a large number of foalings and a knowledge of what is progressive movement of the foetus out of the birth canal, commensurate with the effort being made by the mare.

Foaling cannot be likened to squeezing paste smoothly from a tube. It is, rather, an event of fits and starts. The forelegs appear relatively easily, but the elbows may become lodged on the pelvic brim. This is a time for decision as to whether some assistance should be given to bring the limbs over the brim by pulling on one or other. Another impediment to delivery occurs if the plane of the foal's body is not tilted slightly to one side of the mid-line, in the greatest diameter of the mare's pelvis. This may cause the mare to get up and down or lie on one side and then the other in an attempt to shift the foal's position. The foal's head may sometimes pose a problem if it is lodged between or below the knees as it comes through the mare's bony pelvic hoops. The shoulders and withers are the greatest cross-sectional area of the foetus, and the mare may experience difficulty in pushing them through the birth canal. Finally, the hips of the foal may become lodged against the side of the mare's pelvis.

After Delivery of the Foal

Once the foal is delivered the mare should remain lying down for up to half an hour afterwards. The longer a mare lies down after foaling, the more time there is for her uterus and genital tract to contract, and therefore the less probability that air will enter the uterus. In many cases the placenta is expelled and third-stage completed while the mare is lying down.

Facilities for Foaling

Facilities for foaling vary enormously according to the breed and value of the mares concerned, the layout of the studfarm and the number of mares foaling on the premises. There is no reason, apart from climate, why

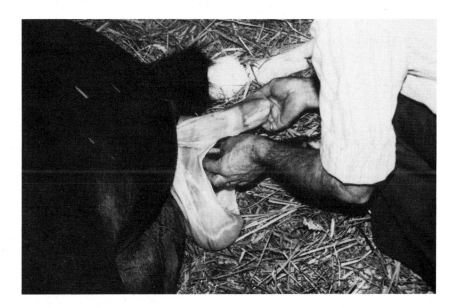

(above, opposite and overleaf) Foaling is an event of 'fits and starts'. The forelegs appear relatively easily but the elbows may become lodged on the pelvic brim. Assistance may be advisable but the decision must be left to the person in charge of foaling; before any such assistance is given it is essential to check that the foal is lying correctly with the limbs and head presented. The minimal force necessary should be applied, always in the direction in which the foal is naturally delivered

mares should foal indoors. In the Southern Hemisphere, where the climate is conducive to mares lying out day and night, mares are successfully observed with lights placed high to the side of the paddock. Of course, mares are quite capable of foaling unobserved and outside; the great majority of mares in the world do just that. It is only when the value of the individual is sufficiently high that the expense of continual observation of all foalings is justified to ensure that assistance can be given to the small percentage (probably five to ten per cent) of mares that require it.

The foaling loose box should be reasonably large, 14ft x 14ft (4.27m x 4.27m) or more. It should have a vantage point through a window or grille enabling the observer to see all parts, and a doorway preferably with a sliding door or one that opens outwards in case the mare should come to lie against it. Some stud managements favour a half-door inside the main doorway. Light fittings in the box should give adequate illumination into all four corners. It is important that the floor of the box be as slip-proof as

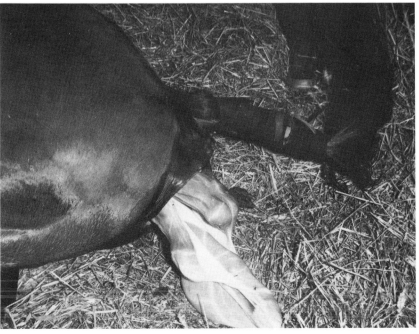

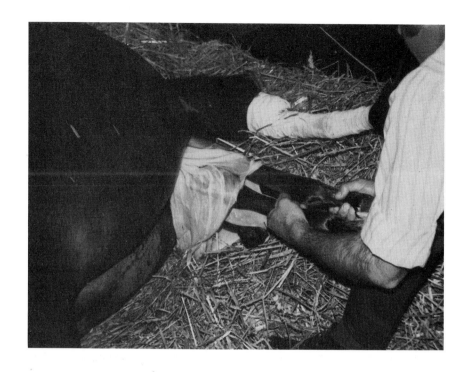

possible, using chalk, striated concrete, rubber or other suitable material. Drainage must be adequate, and the box should be well insulated, especially the roof, and have adequate ventilation. Draughts should where possible be prevented, but the most important aim must be to avoid creating a dusty atmosphere.

The foaling box should have the following facilities: an adjacent room or passage with adequate and accessible means of obtaining hot and cold water, a wash basin for hands, and a sink where the preparation of milk substitutes, etc may be prepared cleanly. A separate area should be available for washing dirty buckets and utensils, such as those contaminated with blood, faeces or meconium. It should also be possible for vets and stud personnel to wash their boots in running water. Cleanliness and hygiene may be achieved with inadequate facilities, but experience suggests that the adequacy of those that are available influences the standards of hygiene practised.

Those responsible for planning facilities in a range of foaling boxes should consider the advantages of having adequate lighting in each loose box and in the passageway or adjoining rooms. One loose box, at least, should be provided with radiant heat lamps of sufficient power to raise the air temperature of the loose box to temperatures approaching 77°F (25°C). Earthed power-points should be placed outside the loose boxes at convenient levels to provide ready access for extra lighting, radiographic and other veterinary equipment.

The 'sitting-up' room should contain a cupboard of suitable size to keep such items as a thermometer, scissors, bandages, medicaments, disinfectant and utensils for preparing feeds of artificial milk for orphan or sick foals. The cupboard should be placed in such a position that it can be approached with ease and its contents kept under review to ensure their availability in emergencies and on other occasions. An oxygen cylinder should also be kept for use in cases of ill foals. We shall return to this necessary item in chapter 12.

The foaling box should be disinfected thoroughly after each foaling, at a time when the mare is moved to another part of the studfarm. All dried excreta and other material should be scrubbed from the walls, floor and mangers. Attention should be paid to the water manger, and the water should be changed. All bedding and fodder should also be removed and a fresh supply introduced. The box should, if practical, be kept empty for two or three weeks after foaling before another pregnant mare is housed there. Unfortunately, this precaution against the spread of infection is rarely practical, except where mares are housed in an American-type range of

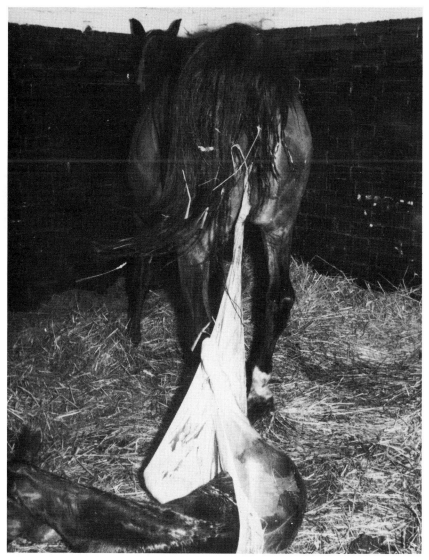

As the mare gets to her feet after delivering her foal the cord ruptures and air may be heard as it is sucked into the uterus through a distended birth canal

The outside of a foaling unit adapted from existing buildings by constructing a passage behind the loose boxes at Ashley Heath Stud, Newmarket

loose boxes with a common overhead area of ventilation and a passage from which the nightwatchman can observe the mares. This is the optimal arrangement for foaling boxes because it allows for long-term stabling of individuals rather than for mares to follow one another in succession through a centralised foaling box. The traditional arrangement on studfarms has been to have two foaling boxes adjacent to a sitting-up room, but nowadays most studfarms that have to cater for large numbers of foaling mares usually have some system to provide a greater number serviced by a passageway. The need for careful control and disinfection of foaling boxes is emphasised by the risks involved of foaling a mare suffering from the rhinopneumonitis virus or even the equine infectious anaemia virus, and then having another pregnant mare replace the infected mare for foaling in the same box. The same principle applies to other infectious agents, although these may not be as dangerous as the two examples quoted.

The air entering the genital tract of the mare when she gets to her feet after foaling carries dust laden with fungus and other microbes which may irritate and cause infection of the uterus at the foal heat or, more seriously, over a longer period. Microbe-laden dust will also be inhaled by the newborn foal at a time when its immune resistance is at its lowest, and

193

Part of the 'sitting-up' passage at Ashley Heath Stud showing a table for veterinary and other use, a hot-air blower, a three-point electric plug for X-ray equipment, electric blankets or other equipment needed in emergencies

may therefore be the cause of infection of the lungs and other organs. Good-quality straw is essential as bedding for foaling mares and, in particular, it should not be shaken out from the bale within the loose box. After a mare has foaled it is customary to remove dung and wet straw and to replace the bedding with fresh, dry straw. This must be placed in the loose box after being shaken outside, rather than inside, the loose box. The mare should be offered a feed after foaling, preferably containing some wet bran and green food. Hay and water should be available at all times during and after foaling.

SOME SPECIFIC DUTIES OF STUDMEN AT FOALING

The studman attending foaling has the prime responsibility of diagnosing if foaling is abnormal in any respect, and of taking appropriate action or calling for veterinary assistance. What has already been written above is intended as an aid to the understanding of the foaling process; here we may confine ourselves to some definitive actions, commonly taken by studmen during foaling.

'Stitched' Mares

Mares that have had their vulva sutured (Caslicked: stitched) must be cut before second-stage can proceed. If this simple precaution is not taken the foal will cause a jagged tear that may cause problems in healing and deform the perineum with scar tissue. A straight cut, using scissors with 6in (15cm) blades, should be made to the level of the roof of the vagina. Both the extent of the cut and the fact that the mare has been previously stitched may be tested by running the forefinger upwards between the vulval lips. If the mare is 'unstitched' the finger will not catch on any fold of skin. The best time to make the cut is when the mare has broken water just before the forefeet appear at the vulval lips, which at this stage generally have little or no feeling. Many studmen like to make the cut while the mare is lying down, but some mares are unco-operative and get to their feet on approach. These must be cut in the standing position with due precautions against being kicked.

If, for any reason, this surgical procedure cannot be satisfactorily completed at foaling it may be performed by the vet some hours or days prior to foaling, using a local anaesthetic. Resuturing the wound produced by cutting may be performed immediately after foaling, usually without the need for a local anaesthetic, because the vulval lips are still numb at

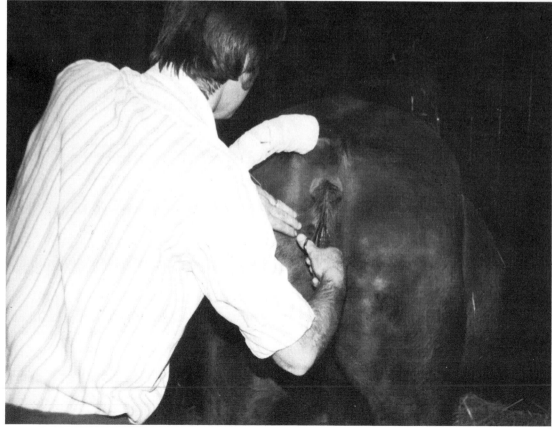

this stage. Later, local anaesthesia is essential. Resuturing is a skilled operation and requires veterinary supervision.

Feeling for the Foetal Foal

The studman, having noted the onset of second-stage, should examine the foetal foal as it enters the birth canal. This can be done with the mare standing or lying down. In the standing position the foal is pulled back into the abdomen by the force of gravity, and there is thus more room to work, although the operator's arm has to be inserted more deeply than is the case with the mare lying down. In this situation the foal is pushed into the birth canal; the nose and forelegs can be more readily felt by the operator with a hand in the mare's vagina. The attendant in charge of foaling should sponge down the mare's perineum, and then, wearing a glove or having washed his/her arm in warm, soapy water, feel for the presence of the foal's muzzle and two forelegs. The forelegs may be identified by feeling the feet and then following the limbs to the fetlock and next to the knee. Having identified the muzzle and the forelegs the operator must judge whether or not the foal is in the correct position and not upside down. Practice and experience are the two essentials in learning the correct interpretation of this important manoeuvre. If the foal's appendages are incorrectly aligned it is necessary to take action to correct them. In these circumstances the operator must first make a correct diagnosis. The mare should be made to stand during any attempt to correct a postural malalignment. In this position the foal, being pulled into the abdomen by gravity, is more readily manipulated than if the mare is lying down. Every person faced with a problem must judge whether or not they are sufficiently experienced to diagnose and correct the abnormality. An unavailing attempt to correct a malalignment may make matters worse; and the delay may prove fatal to the foal. If in doubt it is much better to call the vet than to risk making a mistake. The vet will not be upset if he should arrive and find that the foal has been born and all is well; he may quite rightly be annoyed if he has not been given the chance to save the

(opposite, top) *Mares whose vulvas have been previously 'stitched' (Caslick operation) should be opened before delivery starts. The lying-down position is safer and preferable but sometimes it is necessary to cut the vulva in the standing position* (opposite, bottom); *in this case it is essential to take precautions against being kicked although the vulva at the start of second-stage labour often has little feeling*

foal's life because he has *not* been called in time.

It should be possible to distinguish between the normal presentation and posture of a foetus, being delivered head and front feet first, from abnormal posture (eg a flexed knee, presence of two front feet but no head, presence of head but no front feet) or incorrect presentation (eg feet and hocks palpable but not the head). All these abnormal alignments of the foetal foal are likely to cause delay in delivery. Thus the observer may gain some intuitive sense of their presence before actually feeling them on examination.

Correcting Minor Irregularities

As delivery proceeds, the head may appear to become lodged below the forelimbs or be causing some obstruction in delivery. Gentle manipulation of the head and re-arrangement of the forelimbs may be helpful. A common deviation is for one forelimb to become markedly retarded compared to the other. Again, gentle manipulation of the retarded limb may be helpful in speeding delivery. However, it cannot be overemphasised that care should be taken in re-arranging the appendages to avoid damaging the foetal chest. Pulling vigorously on the forelimbs and extending the elbow and shoulder joints may expose the chest to serious damage as it negotiates the bony prominence of the pelvis, pushed by the enormous power of the mare's expulsive efforts. The operator must be firm but gentle, careful yet decisive in effecting any changes in posture to facilitate delivery. Experience is the most important attribute in these circumstances, but it is useful for the novice to know that even when delivery is prolonged, the foal's lifeline, ie the umbilical cord, remains intact and unmolested as long as the foal's chest has not reached the pelvic outlet. Once the chest is leaving the birth canal, the umbilicus is passing over the pelvic brim and the cord may come under pressure. From this moment onwards, but only then, speed of delivery is important. The cord, at this stage, is being dragged over the pelvic brim and may be squeezed, by the foal's belly, against the bony floor of the birth canal, thus obstructing the blood flow from the placenta and reducing the foal's supply of oxygen. Once the chest is free of the birth canal, the foal obtains oxygen from the atmosphere by breathing; and constricting the umbilical cord does not matter. However, the foal cannot make breathing movements if its chest is still engaged in the mare's vagina, and there is therefore a short but dangerous interval during which the foal is at risk from suffocation. This dangerous moment usually passes without

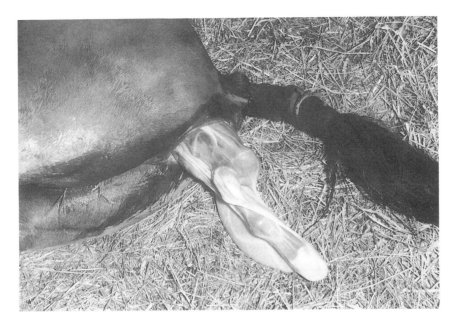

(above) *At this stage the umbilical cord is, normally, in no danger of being squeezed and the foal's lifeline is therefore intact;* (below) *the cord is now being dragged over the pelvic brim and delivery must be completed quickly to allow the foal to start breathing*

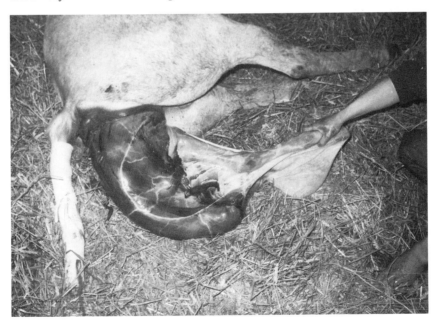

incident, because the chest comes free quite easily, pulled partly by the weight of the foal's foreparts already delivered and partly pushed by the mare's expulsive efforts. In some cases the foal's hips become engaged against the mare's pelvis and, despite straining, she cannot deliver her foal completely. Some assistance is then necessary in the form of gently easing the hind parts out of the birth canal by rotating them slightly one way or the other. This action may be achieved by gently pulling on the forelegs in the direction of the mare's hocks, while a second person grasps the skin on either side of the belly and rotates first one way and then the other. This usually dislodges the hips, which then flop from the birth canal, and delivery is complete.

Preventing Suffocation

Should we remove the amnion from the foal's muzzle? This question is often asked, and the most suitable reply is that a normal healthy foal has no problem in opening the amnion when it arches its neck and strikes out with its front feet at the taking of its first breath. However, a foetal foal that has been affected by illness during pregnancy or which has been subjected to exceptional stress during birth may be unable to achieve the breakout. It is, therefore, at risk to a potentially suffocating situation in which the membrane covers the nostrils or the muzzle is immersed in amniotic fluid as the foal starts breathing. It is good practice to break the amnion and to peel it from the head immediately the foal has been completely delivered, and to raise the foal's head out of the amniotic fluid if it does not do this of its own accord. This action will be unnecessary in most cases, but we have no means of forecasting that a foal may be suffering from illness or is severely stressed by the birth process, before the foal is actually delivered. There are, of course, exceptions, such as if the amnion and amniotic fluid are seen to be stained with meconium, because this is a sign that the foal has suffered some shortage of oxygen before or during birth. It is thus particularly important in such a case to take appropriate action. We must return to the duties of studmen when we come to consider the newborn foal (see page 195).

The Umbilical Cord

It was once the custom of stud grooms to grasp the cord, cut it and tie it as soon as they could when the foal was completely delivered. This practice has largely been discontinued because it was shown that to sever the cord

Foal being delivered by gentle traction of the forelegs, with the head just emerging from the mare's vagina. It may be prudent to break the amnion and lift the foal's head out of the amniotic fluid immediately it is delivered

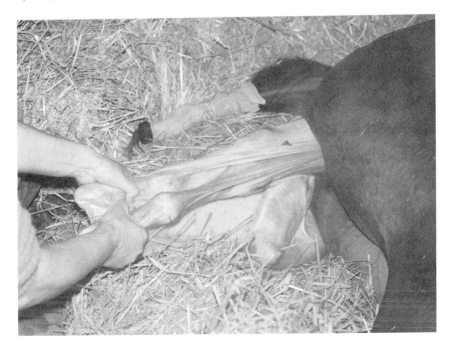

before the foal had started breathing may deprive it of as much as one-third of its total circulatory blood volume. It is, therefore, important to leave the cord to break naturally by one of two ways: either by the foal stretching and pulling it as it struggles in attempts to get to its feet; or by the mare rising to her feet. The cord becomes white and brittle as the blood circulation passing through it, between the foal and the placenta, diminishes. The cord breaks at a natural point close to the belly of the foal. It is usual to dress the stump with antibiotic powder, although this is not essential.

If the cord ruptures in the manner described, the blood vessels become sealed and there is no need to place a ligature. If bleeding should occur, it is only necessary, in most cases, to pinch the stump between the thumb and finger for the bleeding to stop. Should the bleeding continue, however, a ligature of tape may be placed firmly around the stump. Abnormally thick cords may have to be ruptured by hand. In this case the umbilicus is supported with one hand while the operator pulls sharply on the cord with the other.

After delivery, the foal struggles in attempts to get to its feet and in so doing moves away from the mare stretching the umbilical cord

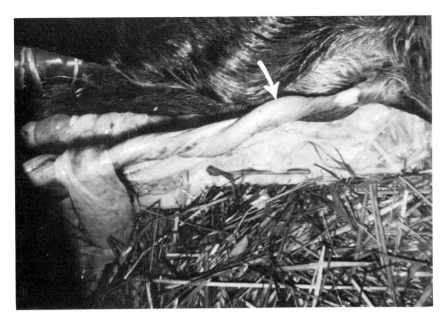

(above) *The blood circulating through the cord (the large vessel arrowed is the vein carrying blood from the placenta to the foal) diminishes as the cord becomes stretched;* (below) *a white area develops close to the umbilicus and the cord ruptures leaving a stump which may be dressed with antibiotic or anti-septic powder*

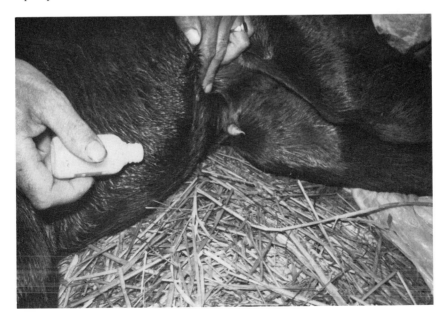

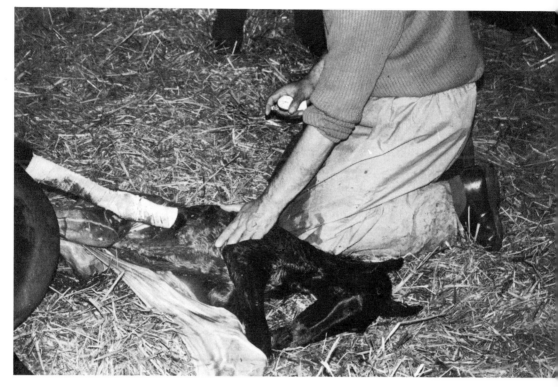

Recording the foal's heart beat on delivery tells us something of the state of the foal at birth

Foal's Heart Rate at Birth

It is good practice to record the foal's heart beat at birth. The normal rate is between forty and eighty beats per minute. An abnormally slow or fast rate may indicate that the foal has suffered birth asphyxia (see page 289).

Drying a Foal's Coat

Studgrooms often dry the foal's coat which is saturated with amniotic fluid. This practice is not essential but it may help in cold weather if the foal is suffering from the effects of delivery or from an abnormal pregnancy. The practice not only reduces the amount of heat lost by the foal through evaporation but stimulates it in general and accentuates the reflex activities of breathing and movement which are essential to survival. However, it should not be thought that the procedure is necessary in any but a small minority of cases.

The thermoneutral range is the range of air temperatures at which the foal's metabolic rate (heat production) is at a minimum. Within this range, the foal will have a normal rectal temperature (about 38.5°C or 101.3°F). When air temperatures increase, a foal will begin to sweat in order to dissipate heat. However, when air temperatures decline, the foal will need to increase its metabolic rate and subsequently will start to shiver; the temperature at which this increase commences (the lower critical temperature) is about 20°C for pony foals, and 10 to 15°C for Thoroughbred foals. Normally foals can thermoregulate very effectively, but healthy foals that are wet at birth, and sick foals, are susceptible to cold air temperatures and should be maintained within their thermoneutral range.

7

THE NEWBORN AND OLDER FOAL

The newly born foal is none other than the foetal foal that we met in chapter 6. But now we can see and appreciate it, examine and study it directly, subject it to tests and diagnose its ailments.

It is fascinating to contemplate the differences and the similarities between the existence of the foetal and the newborn foal. We have seen how, in the uterus, the foetal foal's well-being depended on the placenta as an organ of exchange of gases, nourishment and waste. Its lifeline was the cord, with its vessels carrying blood to and fro. Birth separates the foal from its protective niche, and severs its connection with the placental membrane. The most immediate challenge of its new environment is the need for oxygen. Breathing is thus the first and most necessary step that the newborn foal must take in its struggle for survival. As its lungs fill with air for the first time, oxygen levels in the blood rise enormously, thus meeting the increased demand necessitated by the vastly higher activity of the body tissues. The muscles, in particular, have to meet the efforts required to lift the foal to its feet; and energy is required to enable the foal to gallop with its mother. The newborn foal is an athlete within a few hours of its birth.

Some of the challenges encountered by the newborn foal are similar to those experienced in the uterus, but of a greater degree. For example, the heart has been pumping blood throughout the body and through the placenta for some forty out of forty-four weeks of pregnancy. However, the workload of the pump is dramatically increased during and after birth. In the uterus, the foetal foal is relatively inactive; and any movement is cushioned by the surrounding fluid and the mare's abdominal contents. Outside the uterus, however, the foal is much more active and each movement is made against the pull of gravity. The energy demands are therefore correspondingly increased and have to be sustained by similar increases in metabolic activity and cardiovascular (heart and circulatory) performance. A summary comparing intra- and extra-uterine life is shown in Table C.

TABLE C Life inside compared with outside the uterus

Activity	Foetal	Newborn
Respiration (Breathing) (Exchange of gases, oxygen and carbon dioxide)	between mare's blood stream in uterus and foetal blood stream in placenta; lungs contain fluid, not air, and do not function as organ of exchange	between air and the foal's blood stream in the lungs; air sacs of lungs contain air, and organ acts as site of gaseous exchange between air and blood stream
Muscular movement and tone	virtually free of gravity; periodic movements but not strenuous	against the pull of gravity; demands for athletic performance, getting up and down etc
Nervous	need for co-ordinating foetal movements, little sensory input*	co-ordinating complex movements and dealing with substantial sensory input
Light, sound and touch	not subjected to these stimuli to any, or to a very reduced, extent due to being enclosed in the mare's abdomen	subjected to considerable outside stimuli to which it is necessary for the foal to respond in a purposeful and controlled manner, eg moving towards sounds representing useful experiences and away from those indicating harm or danger
Alimentary (Digestion)	nourishment received through placenta from maternal into foetal blood stream; digestive processes completed by mare before becoming available to foetus	active search for food necessary; digestive processes completed in gut
Excretion	urine formed by kidneys, stored in bladder and allantoic cavity, entering by way of the urachus; solid material stored in colon, caecum and rectum as meconium	large amounts of urine formed and passed by 'staling'; faeces formed in large intestines and passed through anus to outside

Metabolic	liver and glands produce hormones and enzymes to meet relatively low demands of metabolic activity	high level of metabolic activity met by increased output by glands and liver
Cardiovascular status	heart pumps blood through the body and the placenta in a relatively low-pressure system	heart pumps blood through relatively high-pressure system; placenta replaced in system by the lungs
Body temperature	body temperature maintained by surrounding tissues and organs of mare; foetus has little need therefore to 'burn' sugars to maintain body temperature	loses heat from body by radiation, convection, evaporation and conduction; necessity of burning sugars to maintain body temperature against surrounding environmental conditions
Behaviour and socialisation	not required	essential bonds must be formed between mare and foal, and between other members of herd

*sensory input = messages received by feelings such as of pain, noise, light, heat

THE NEWBORN PERIOD AND ADAPTATION

The Neonatal Period

The neonatal period comprises the first four days following birth. During this time the major adjustments to life outside the uterus are completed. It is also the period in which symptoms of disease and conditions peculiar to the newborn first appear. But it should be emphasised that this definition is only for convenience, providing us with an order for description and understanding.

The manner in which the foal's body responds to meet the challenge of its new environment is termed adaptation or adjustment. Failure is termed maladjustment or maladaptation. Normal adaptation can be interpreted by the foal's behaviour: for example, breathing within thirty seconds to a minute of being delivered, turning onto the brisket within five minutes, standing within ninety minutes, and sucking from the mare's udder within 200 minutes. The landmarks of normal adaptation are described in Table D.

TABLE D Landmarks of normal adaptation in the first four days of life

Age from complete delivery

1 minute	breathing, rapid movements of chest and abdomen; heart rate 60–80 beats per minute; temperature 99–100°F (37.5°C)
5 minutes	breathing rhythm fully established; starting to lift head and turn onto brisket, at same time withdrawing hind legs from vagina; shivering; suck reflex present; temperature 99–100°F (37.5°C)
15 minutes	turned onto brisket, front legs stretched in front, hind legs flexed underneath and maybe making attempts to stand; breathing rate 40–60 movements per minute, heart rate 120–160 beats per minute
30–90 minutes	foal stands for the first time; starts search for udder
1 hour 40 minutes – 3 hours 20 minutes	foal sucks from mare for first time and establishes instincts of following; passes meconium
12 hours	instincts of following mare and sucking strongly established; able to get up and down with ease; passing urine and meconium; temperature 100–101°F (38°C); heart rate 80–120 beats per minute, breathing rate 30 per minute at rest
48–72 hours	milk dung appears and meconium completely evacuated

Management of the Newborn Foal

In the last chapter we left the foal at the point in time when it had been delivered by the mare and had just started to breathe. Management should, as suggested for foaling, adopt a watchful attitude; being prepared to stand back and let nature take its course if events are normal, but interfering decisively if not.

Once again, the duty of studmen to interpret the events unfolding before them requires a sound knowledge of the normal. We have already discussed some of the salient features that must be recognised as a mark of normality, such as standing for the first time within ninety minutes and sucking within 200 minutes of delivery. A foal that fails to achieve these objectives within these time limits is not necessarily abnormal but such a foal must be viewed with some suspicion, and it may be useful to seek veterinary advice. We shall find, in discussing diseases of the newborn foal, how very important it is to act quickly and in time to prevent the condition deteriorating. Young foals do not have the resistance of older horses; and a few hours' delay may prove serious, even fatal.

Behaviour as an Indicator of Health

The behaviour of foals can be an all-important indicator of health or ill-health; the way a foal stands, gets up and down or lies on the ground tells a story of some importance. For example, foals normally rise to their feet in an adult manner, ie by stretching out their front legs and first lifting their front parts followed by their hind parts; they lie down in the opposite sequence. They may have some difficulty during the first hour or two after they are born, falling about and overbalancing. However, once they have stood successfully for the first time, their movements become increasingly well co-ordinated, and standing is achieved in a smooth, sequential manner. A foal of more than age three hours which half rises to its feet and

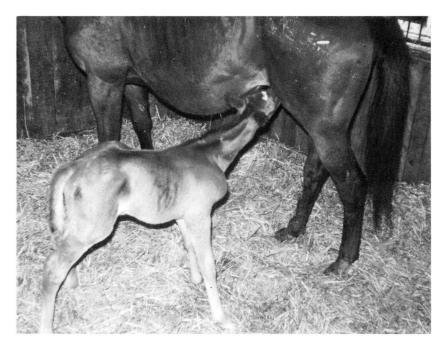

The manner and ease with which a foal finds the mare's udder and sucks is a good indicator of health; here the foal is holding the sucking position

Healthy foals lie in a relaxed manner sometimes outstretched (above) *or* (below) *on their brisket*

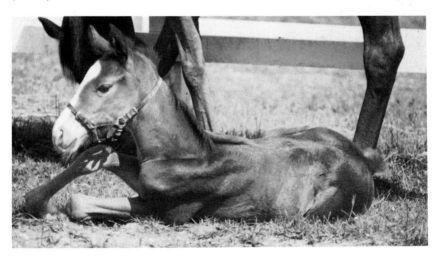

then falls over, or which does not appear to have the strength to stand, may be suffering from infectious or other disease.

Another indicator of weakness or strength is the way in which the foal seeks the mare's udder, takes hold of the teats and holds the sucking position. A healthy foal has little difficulty in finding the teats and it will certainly persist in its search if it does not find them immediately it reaches the mammary region. A weak foal, however, may experience considerable difficulty in finding and holding the teats for any length of time. The presence of the foal's muzzle round the udder stimulates the mare to let down her milk, and a foal that is not sucking properly may have its nose and muzzle covered with milk as it streams from the mare's udder unconsumed.

Foals spend about one to two minutes per sucking bout, which recurs three to four times an hour in the first few days of life, but the frequency and time spent at the udder tends to decrease with age. The state of the udder and the response of the foal in approaching and sucking from it are a cardinal indicator of the foal's well-being or otherwise. A foal that has been lying down and is made to get up should approach the udder and show interest in sucking as soon as it is left undisturbed. A sick foal may show no interest at all or merely nuzzle round the udder without taking

Foals in pain may roll or lie in 'awkward' positions

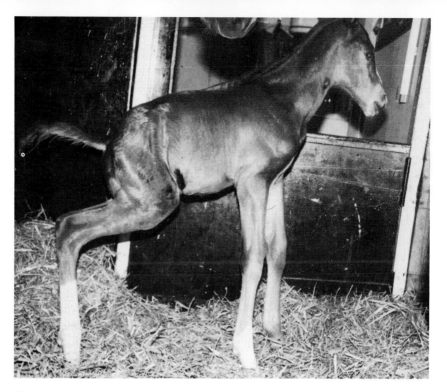

Foal crouching and tail raised in an effort to pass urine or meconium

hold of the teats and sucking, depending on the nature and severity of its illness. Studmen should make themselves familiar with the various states of the udder itself because this can indirectly provide clues as to the well-being of the foal. A flat, slack udder with a moist surface round the teats indicates that the foal has recently sucked and fed well, whereas dry teats denote that the foal has not sucked within the last half-hour or so. A full udder may be distinguished by sight or feel but it should not be confused with the large fleshy udder possessed by some individuals. The difference may be readily confirmed by feeling the tension in an udder full of milk and by squeezing the teats to determine the quantity present. Milk, in a full udder, may be seen as a drop at the end of the teats, or dripping or streaming from the teats.

The way a foal lies when resting provides further indication of its well-being. Normally a foal lies outstretched on its side with its legs resting naturally at right angles to its body and slightly flexed. When on its brisket the foal's knees are flexed and the hind legs drawn beneath the body. It may turn its head towards its flanks but this action is not usually exaggerated or repeated as is the case with a horse suffering from colic. Foals in pain may roll or lie in awkward positions, looking back at their flanks or lying outstretched with their heads turned back. The attitude of standing may also be significant as a sign of ill-health. But the normal must be studied to appreciate the abnormal.

214

Meconium

Squatting, straddling and straining are common stances to be seen in the first three or four days after birth. During this period they are probably exaggerated forms of behaviour associated with passing meconium.

Meconium is a dark brown, black or greenish dung with a somewhat rubbery consistency. It is stored in the rectum, colon and caecum during foetal life and expelled after birth as the process of taking of food through the mouth and digestion in the intestines begins for the first time. Meconium is expelled by the normal process of defecation, but the foal may have difficulty in pushing large lumps out of its rectum over the relatively small exit formed by the bony pelvic girdle. This problem is particularly marked in colt foals, where the outlet tends to be smaller than in females. However, the meconium extends a long way back (in the intestines) away from the pelvis. To void this long column of faecal material is not, therefore, simply a matter of squeezing it through the pelvic outlet. It is also necessary for the gut wall to propel the meconium in a backwards direction towards the rectum and anus. This is achieved by peristalsis, ie waves of muscular contraction passing down the tube formed by the walls of the small and large gut. Stretching of the walls by gas or accumulated faeces is painful and gives rise to signs of colic. Undue stretching of the gut wall diminishes its capacity to contract in peristaltic waves and causes painful spasms. Passing meconium is based on a process of 'push' by the gut from behind to overcome the obstacle to its evacuation formed by the constriction in the rectum at the pelvic outlet. Meconium retention is a condition in which there is diminished ability of the intestines to propel lumps of meconium.

Enemas

In foetal life meconium is never passed except during abnormal circumstances (usually associated with an acute shortage of oxygen). However, it should be evacuated over the first three days after foaling. This is usually accomplished with little difficulty because the lumps are lubricated by a slimy mucous covering, though it may be helpful to administer additional lubricants and enemas to assist defecation. Proprietary fluids containing special substances that lubricate and soften dung pellets in the rectum may also contain compounds with mild laxative effects that promote peristaltic movement in the small colon and rectum. The traditional means of giving enemas of soap and water or injecting quantities of liquid paraffin necessitate the use of rubber tubing which should be soft and have no sharp

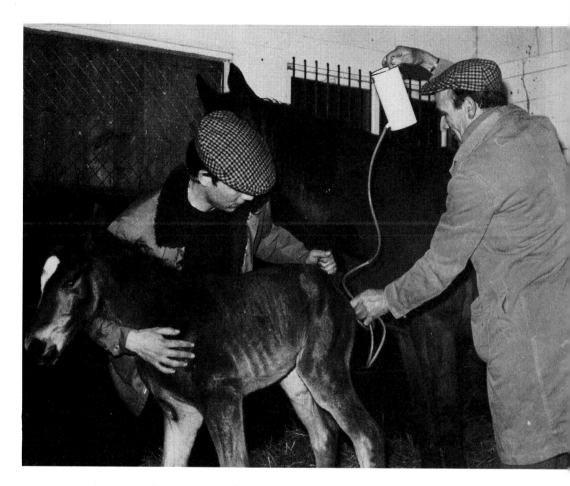

An enema of warm soap and water being administered to a young foal; note the manner in which the foal is being held. It is important to avoid using force on the tail because this bruises the delicate tissues at its base; the tube must be soft and free from any sharp edges

edges that might bruise or tear the anus and rectum. Force should *never* be used to insert the nozzle into the rectum past the accumulated meconium pellets. Further, too much fluid and/or too-frequent enemas tend to cause the rectum to balloon. This may reduce the effectiveness of peristalsis in squeezing the contents of the rectum through the anus, thus causing and not curing the problem. The lubricating effect of oil or soap is helpful in small quantities, but, as already explained, the column of meconium extends well into the colon and out of reach of enemas. In severe cases of retention we must, therefore, try other measures such as administering oil by mouth or through a stomach tube, so that the lubrication works from behind the impacted mass.

There are two schools of thought as to whether the first enema or paraffin lubrication should be administered soon after birth or after the first suck. The disadvantages of administering enemas before, rather than after, the first suck that the foal is less amenable to handling at this stage, and there is more pressure in the abdomen in the lying as opposed to the standing position. Further, an advantage of administering the enema after sucking is that the swallowing of milk, in itself, promotes reflex peristaltic movements of the hind gut leading to defecation. Thus the enema in these circumstances coincides with and is helpful to the gut activity. Colostrum, the mare's first milk, is a powerful laxative which will eventually clear the hind gut of meconium as it passes through the length of the intestines.

Meconium can best be evacuated by means of a gloved forefinger, lubricated with liquid paraffin, inserted through the anus, and the pellets within reach gently withdrawn into the palm of the hand. If the foal strains at this juncture it assists the operation but if, as is often the case, it remains passive, only that meconium lying above the floor of the pelvis can be removed. Meconium that is in the rectum but behind the brim of the pelvis may be felt at the end of the finger but not brought back. However, by removing some of the meconium, some space is provided for an injection of a small amount of liquid paraffin, and the foal may then be left until more meconium passes into reach. Attempts to remove meconium or administer enemas should not be made too frequently for reasons already stated, ie they may cause bruising or ballooning of the anus and rectum.

Handling the Newborn Foal

The handling of a foal before it sucks may make it panic and struggle quite violently, thus adding to the stress it is already suffering as a result of arriving in its new surroundings. The majority of foals nevertheless come to no harm, but those on the borderline between normality and abnormality may succumb as a result of the extra exertion and suffer from convulsions or other similar conditions of maladjustment (see page 303). Newborn foals should be handled in such a way that they are put under stress as little as possible. A quiet sympathetic approach is of the utmost importance. The most successful and safest method of restraint is to place an arm around the front and hind quarters. A second person can perform such tasks as fitting a head collar, administering small amounts of fluid or medicament by mouth, giving a hypodermic injection into the hind quarters, into the muscles behind the shoulder or into the neck, or

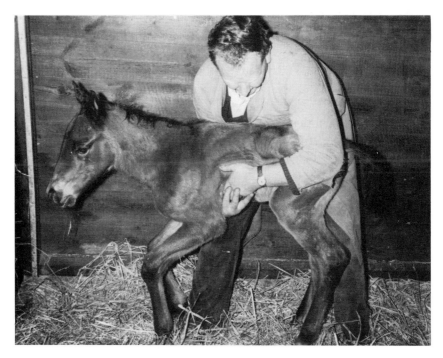

An experienced studgroom demonstrating how not *to lift a young foal; this method places too much pressure on the chest*

administering an enema. The tail is a useful appendage to restrain a foal, but if we handle this part too vigorously we are sure to bruise the delicate tissues at its base causing swelling (oedema) round the anus. Foals should never be lifted by their tail, nor should their weight be taken on the keel of the chest because this action may fracture the ribs. Young foals usually co-operate and settle if carried with the arms around the shoulders and hind quarters, and from this position we are able to lay them down. Restraint on the ground is best achieved by raising the head and turning the muzzle uppermost. There is generally no need to do anything more than soothe the foal by kind words and gentle stroking of its neck; and certainly holding the legs together is unnecessary and counterproductive, as it causes the foal to struggle in attempts to free its limbs. The position illustrated on page 318 is the one favoured as a basis for the control of foals suffering from convulsions and other behavioural disturbances of the newborn period.

Instinct and Learning

Much of animal behaviour is geared to a pattern typical of the species. Some aspects of behaviour are inborn and others are acquired; but both are subject to positive and negative influences of the environment. Thus if we analyse behaviour we cannot necessarily predict how a horse will behave in any particular circumstance. However, in the newborn period we can distinguish, to a greater degree than in any other period, between inborn instinctive and acquired behaviour. For example, the manner in which a newborn foal rights itself onto its brisket and then positions its front legs for lifting itself to its feet is identical to that employed by an older horse. Thus there must be more instinct than learning in the process, even though learning by trial and error may play some part as the foal tumbles about in its first attempts to stand. It is also presumably instinct which leads it towards the darkened under-surface of the udder and to the smell and feel of the mammary glands where its instincts are rewarded by an adequate supply of 'food'. But, again, trial and error play a role as the foal often goes first to the brisket and sucks from the skin around the stifle. It is by instinct that the foal follows its dam, latching onto the first large moving object that it sees in the first moments of sight. Thus the

Young foals usually settle if held in this manner for weighing them on bathroom scales

The method of laying down a foal

A healthy subject being used to demonstrate the position which might be used to feed the foal artificially through a stomach tube or to practise any other manoeuvre

bond between the foal and the mare is established by a sequence of behavioural instincts, involving reflexes, that enables the foal to stand, walk and gallop, and by which it can reach and follow its dam. Sight, smell and hearing enable it to maintain contact with the mare, to discern the position of her udder, and to reach the teats for the purpose of imbibing milk.

This sequence of behaviour, one stage leading to the next, is essential for survival and involves little in the way of learning. It may easily be disturbed by external influences, such as managerial interference. For example, someone wearing a brown overall, handling the foal before it has established knowledge of its dam, may become the object of the foal's care-seeking behaviour. Foals may become confused during the first hours after being born and fail to develop a strong bond with the mare. In some cases this may be due to brain damage (see page 316), but it seems that the strength of the instinct varies with the individual and it is to those foals with weak instincts that managerial influence can, in itself, do harm. The same variation of strength is found in the maternal instinct, so it is important for management to ensure that no action is taken that may prevent, interfere with or break mare/foal and foal/mare relationships.

Harmful effects may come from movement and contact by people in the loose box, which result in the foal attaching itself to an individual or becoming unsure as to which 'object' it should be following. Even under natural conditions, it is possible for a foal to become confused by an object in the vicinity of foaling. Dr Stephanie Tyler, who studied New Forest ponies, reported one foal which became attached to a tree rather than to its dam. Foals born into loose boxes may attach themselves to mangers or the dark surface of walls in preference to their mother. Some foals are very easily frightened, and handling may produce panic reactions that abolish the normal instincts temporarily or permanently. Acquired behaviour develops with experience, and in the first hours and days after birth the foal is particularly receptive to learning. The period of imprinting, as this is known, enables the foal, under natural conditions, to strengthen the instinctive bonds that establish its care-seeking behaviour, which will last until weaning and be gradually replaced by social bonds binding it to other members of the herd.

THE OLDER FOAL: TO WEANING AND BEYOND

Once the foal has reached the age of about five days its progress towards a life independent of its mother, and the fulfilment of its destiny as a fully

Young foals play with one another and thereby learn herd instincts; they also indulge in mutual grooming activity with each other or with their respective dams

mature, reproductive individual, proceeds at a rather less dramatic pace than that experienced in the first few days after foaling. Nonetheless, the process of development is continuous and subject to interference by environmental, managerial and microbial influences.

If we follow the foal through these stages of development we must be impressed by the most obvious change, namely the growth to some ten times the birth weight by the time the individual is two or three years old. The majority of this growth occurs within two years; and half of it within one year of birth. The increase in stature is dependent particularly on the growth of bone and, to a lesser extent, a commensurate increase in muscle.

Whole books could be written on the processes of growth relating to each system of the body. Here we must confine the narrative to those elements essential to understanding the practical management of healthy horses in their development from birth to maturity. In practice, if the final results are good we hardly stop to consider why it is we have achieved the optimal endpoint. It is only when we fail to achieve these results that we call into question our methods of management and husbandry.

An understanding of the behavioural patterns of horses at all ages is as important to studmen and others who are responsible for rearing horses as it is to vets for the purpose of diagnosis. Knowledge of normal behaviour provides an insight to signs of ill-health; in many instances, the knowledge is a helpful guide to optimal managerial methods.

The young foal spends much time close to its dam and indulges in frequent short bouts of sucking; as it gets older, it spends more time at greater distances from its dam, sucks less frequently and for shorter periods. Foals start to play with one another during the second week after birth, and their play bouts become more frequent, vigorous and intricate with age. Play is an essential part of herd behaviour, developing the social bonds on which the herd depends for its cohesive unity and reproductive capacity. Play between yearling colts is a necessary preliminary to stallion behaviour, but studmen preparing for the yearling sales do not see it in this light and, to avoid injury, separate the colts from each other in the months leading up to the sales. The same spirit of antagonism develops in mares towards their foals as the natural weaning period approaches. This occurs just before the mare is due to foal, though a barren mare may allow her foal to suck beyond a year. However, weaning in the natural state is a co-operative process between mare and foal. The mare becomes increasingly resentful of her foal's approach as her milk diminishes; and on the arrival of her new foal she will be actively antagonistic. The foal, at the same time, is becoming increasingly independent of its dam and is enjoying alternative fluid and solid nourishment. Thus, weaning is an event of mutual agreement, effected with the minimum of stress for either individual.

Weaning

It is not practical for stud management to depend entirely on the natural means of weaning; it would be uneconomic and very difficult to organise in the context of arranged matings and the movement of individuals from one farm to another. Most authorities recommend weaning when the foal is about five to six months old. There is no particular reason for this preference except that it is based on the idea that the foal at foot may compete with the energy demands of the foetus by taking milk when foetal growth is becoming most marked, ie during the last half of pregnancy. However, there is no evidence that the foetus is harmed by late weaning, provided the mare is being adequately fed. Another consideration is that weaning at age five to six months is convenient for the preparation for public auction in the winter; and the task can be completed before the following breeding season, when staff are fully occupied with foaling and mating programmes.

Method 1 There are two basic methods of artificial weaning. The first method involves the removal of the foal from the mare on a selected date, confining it to a loose box and placing the mare out of earshot. The foal is kept in the loose box for four or five days and then put in a paddock with other foals weaned on the same day. These foals are then run as a group, returning to their loose boxes at night. This method is probably the most stressful because it isolates the individual for long periods, the very opposite of the social companionship that the foal has established under herd conditions. However, isolation is a way of life for older horses who are stabled on their own, and they seem to adapt well to the system. The stress of this sudden severance of contact with its dam and with the herd relationship is considerable. It is at this time that some individuals start those annoying habits found later in life of weaving, crib biting, walking the box and fretting. However, the majority of foals acclimatise to the new situation fairly quickly, although they may temporarily lose weight in the process.

Method 2 The second method of weaning aims to simulate nature's way. A group of mares and their foals are run in a paddock for several weeks so that the social bonds are built up between the members of the group. On

The method of group weaning practised at Cheveley Park Stud, Newmarket: (above) mares and foals being turned onto paddocks in the morning; (below) one of the group being weaned; (opposite) all the foals have been weaned with one mare only remaining

the appointed day of weaning one or two mares are removed from the group and their foals left in the paddock with the remaining mares and foals. The weaned foals may come to the boundary fence from time to time, and may whinny for their dam, but they soon return to the herd and do not seem to worry much about their dam's departure. Foals seldom approach other mares to suck from them, and their presence is usually tolerated by the remaining mares whilst the weanlings provide the companionship for the other foals. After a further two or three days another mare or two are taken away from the group, and the weaning process continues until only one mare is left. This mare may be left with the group or herself removed. The mares are taken away from the group at the time of turning out in the morning, if the group is housed at night. In the evening the weaned foals are placed in their loose box on their own or are paired in loose boxes. They are returned to the group in the paddock each morning. If the group is out in the paddock, day and night, weaning is that much more natural because the mares are simply led away at any time of the day or night and the foals left behind with the herd.

There is, of course, a period for all horses when they are placed on their own for the first time. It is a matter of individual judgement and opinion as to whether this should be at the time of weaning in the autumn, when all foals are stabled, or at some later time, perhaps not until the yearling stage has been reached. In Australia and other Southern Hemisphere countries it is common practice to leave one mare among a group of weaned foals to act as a 'school teacher'. Sometimes an old gelding or barren mare may be used for this purpose. Whatever the decision, we must expect some stress when a horse is asked to live on its own for the first time away from the direct companionship of others.

Growth

The most rapid phase of growth is during the first month after foaling, when the foal's height increases by some thirty per cent. A second growth spurt occurs between six and twelve months, and a third following puberty. Foals born in winter months tend to have lower weight and height at birth than those born in spring and summer, according to a study made by scientists at Cornell University. In this investigation of Thoroughbreds, it was found that the discrepancy in weight and height lasted for at least eighteen months. Thus early-born foals tended to be smaller at a comparable age than those born in the late spring and summer. This finding supports the view that the growth of foals is to a large extent programmed during pregnancy. A foal is born with a growth potential that cannot be bettered by the environment it receives as a foal and yearling — although, of course, it can be made worse by poor environmental conditions and other adverse factors. We have yet to explore the relationship between the quality and quantity of a mare's milk and the ultimate size of her foal. But it is clear that some foals are disappointingly small at nine months despite an adequate diet and good managerial conditions. It may be presumed that these individuals have an inherent incapacity for growth to a size that those responsible for their care would expect. Smallness may be the result of some disease process, such as infection, or the provision of diets with an imbalance or deficiency of minerals, vitamins or protein. However, it may be that these factors play a more significant part in pregnancy, by influencing the foetal foal, than they do once the foal is born. More investigations are required on this fascinating subject which is so important to the fulfilment of the hopes of breeders.

Bones grow through the activity of the growth plates, structures to be found at the end of the long bone, eg the cannon bone, radius and tarsel bones of the fore and hind limbs respectively. The growth plate (see fig 26) consists of cartilage (soft bone) which becomes calcified and changed into new bone, thus adding to the column already present. In this way the bone grows in length and the horse in height.

The growth plates change at various ages depending on their site in the skeleton. For example, the growth plate in the pastern bones closes (stops growing) at 6 to 9 months, those at the lower end of the cannon bones at 9 to 12 months, and those at the lower end of the radius (just above the knee) and of the tarsus (just above the hock) close at 24 to 30 months of age.

It is customary for people to talk of the growth plate being open or closed according to the activity of growth. However, an open plate does not necessarily indicate a relative degree of immaturity, only that the bone in question is still capable of growth. Inflammation of the growth plate is known as epiphysitis (see page 324), derived from the term epiphysis or 'end of the bone'.

Feeding the Young Foal and Yearling

There are many learned works and recommendations on the nutritional requirement of horses, but perhaps the most authoritative is the *National Research Council Recommendations on Nutrient Requirements of Horses* (published by the National Academy of Sciences, Washington DC). The

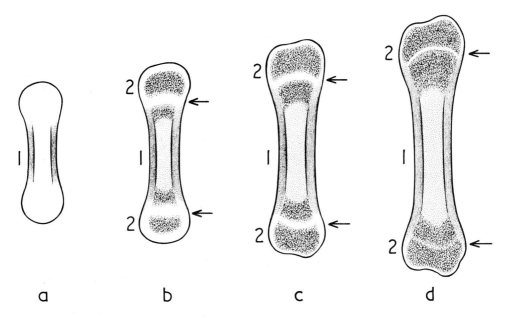

Fig 26 A long bone growing from the foetal size (a) to maturity (d). The shaft of the bone (1) grows at either end (epiphysis-2); the epiphysis consists of a centre of ossification and a growth plate (arrowed) which closes when the bone ceases to grow

recommendations for weaning foals are 16 per cent crude protein, 0.8 per cent calcium and 0.55 per cent phosphorus; for yearlings, 12 per cent crude protein, 0.5 per cent calcium and 0.35 per cent phosphorus. Unfortunately in practice it is rarely possible to regulate the diet to anything approaching such an exact analysis because there are too many variables, such as the quality of the mare's milk, pasture and available feed stocks. Further, there are enormous variations in managerial regimens involving hours at pasture and/or in stables, environmental temperatures, protection against prevailing climatic conditions, and availability of various types of fodder. It is generally accepted nowadays that creep feeding by arranging suitable access to a manger in the stable or in the paddock is of benefit. Dried milk pellets are the fodder of choice offered *ad lib* to foals at foot. After weaning, individual feeds are presented, and dried milk may be incorporated in the corn. It is important that the foal and yearling have access to hay of excellent quality to provide the necessary fibre on which the intestinal tract depends for normal development. Further advice on feeding is presented in chapter 13.

Artificial Diets

Foals that have lost their dam during birth, or at a later stage through accident or disease, may have to be maintained on artificial diets. In terms of nutrition, it is ideal if a foal can be fostered to another mare (see below), although this is not always practical or successful. Therefore, the foal must be provided with an alternative source of milk. Cow's or goat's milk may be used successfully to feed foals (sometimes directly if one can find an obliging cow or nanny goat!) although their milk does not contain as much carbohydrate as mare's milk. Commercial mare's milk replacer, in conjunction with concentrates and roughage, may also be used success- fully, although the use of some formulations has been associated with gastrointestinal problems in the foal.

Some basic recommendations on feeding the orphan foal have been compiled from different authorities and are described below:
(a) It is essential to provide foals with colostrum within the first twelve hours after birth (see page 217).
(b) Diet changes should be made slowly (over 24–48 hours).
(c) All feed should be handled in the most hygienic manner.
(d) Processing the protein in milk substitutes should not be allowed to cause a loss of essential amino-acids.
(e) An excess of liquid feed in any one meal (see below) should be avoided because this can cause a number of complications, for example the growth of pathogenic microbes in the gut or gastric reflux and aspira- tion resulting in aspiration pneumonia.
(f) Bucket feeding is recommended to reduce the possibility of the foal bonding to people, although initially a teat-and-bottle may be preferred by the foal.

Regarding feed volumes for foals, a little, often, is the best approach. Daily milk consumption in healthy foals initially is about 10–15% body- weight, quickly increasing to 20% or more bodyweight. For a foal weighing 50kg this is equivalent to about 12 litres/day (500 ml/h) or 150kcal/kg bodyweight/day. A feed frequency of initially about every 2 hours is recommended, decreasing gradually over several weeks until free choice intake can be implemented. These figures will vary between indi- viduals.

The foal should be encouraged to eat solid feed from an early age. By age two months a foal should be consuming about 1lb (0.45kg) of concen- trates per day; a recommended creep feed supplement consists of 40 per cent cracked corn, 20 per cent soya bean meal, 23 per cent whole rolled oats, 0.5 per cent brewers' yeast, 3 per cent molasses, 1 per cent dicalcium phosphate, 1 per cent limestone, 1 per cent salt, 0.5 per cent vitamin and mineral mix.

The ideal method of dealing with an orphan foal is to provide it with a foster mother. There are several methods of introducing the orphan to the foster mare. These include:

1 dressing the foal in the skin of the foster mare's dead foal or in its amnion. The disadvantages of this method are lack of hygiene, risk of disease, and the possibility that the skin may make the wearer too warm;
2 instilling an odiferous ointment into the mare's nostrils and restraining her with a twitch or administering a tranquilliser and blindfolding the mare;
3 confining the mare in a solid-sided wooden crush with an opening near the mammary region so that the foal can suck without coming into harm's way.

Most of these methods depend on trial and error, and their main objective must be to avoid getting the foal injured by the mare in the process. Readers who have had no experience of introducing orphan foals to foster mares should obtain veterinary advice or depend on the experience of those who have had prior involvement.

Handling Foals and Yearlings

The training of horses is beyond the scope of this book. However, in preparation for work, handling on the studfarm is a most important intro-duction to the individual's acceptance of human methods. Placing a head collar on the newborn foal starts the process which culminates in 'breaking' the yearling. It is also the basis of teaching the foal to lead.

Farriery

The foal is born with a soft, golden, hoof undersurface consisting of overgrown horn. There has been no friction in the uterus to rub this off, but after birth it soon disappears as the foal walks on the firm surface of the ground. Expert farriery should start at about two months of age to maintain the correct shape of the foot and to counteract any deviations from uneven wear; it also helps to accustom the foal to this type of handling.

Yearlings at Derisley Wood Stud, Newmarket, being broken to tack

Expert handling of young foals is an important introduction to the more serious handling necessary for breaking and riding

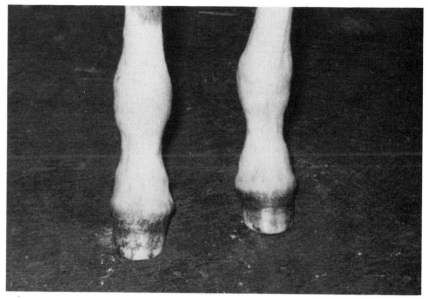

Farriery should start at about two months of age

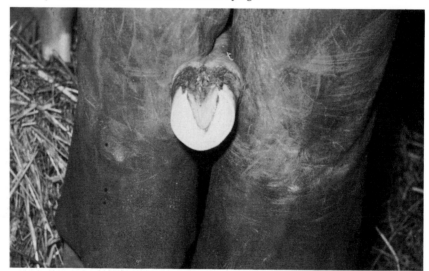

The foot must be kept in shape and excess horn pared away to prevent the toe from becoming too long or the sole 'one-sided'. Here the frog is seen to be central, as is necessary

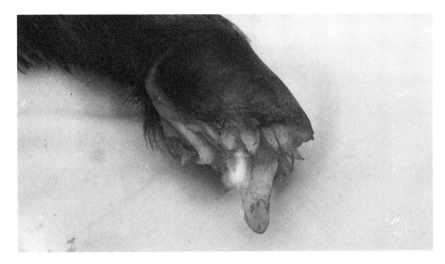

The 'golden hoof' of the newborn foal: this soon disappears as the feet meet
mother earth instead of the soft surroundings of the maternal uterus during
pregnancy

The farrier at work; the foal is being held against the wall and cannot go
forward because of the manger in front of it

PART III

ABNORMALITIES
AND DISEASE

8

THE VETERINARIAN'S
ROLE ON THE STUDFARM

The duties of the vet on a studfarm may be discussed in terms of managerial and veterinary functions. The managerial element is best illustrated by a short historical account.

HISTORY OF VETERINARY INVOLVEMENT

Before the Second World War the breeding of horses depended on a simple knowledge of the oestrous cycle, namely that when a mare shows heat she should be mated and that two to three weeks after she has not shown heat at the expected time she might be considered as being in foal. The custom was, therefore, to tease mares with a male horse and to observe for signs of oestrus and dioestrus. When the mare was in heat the studgroom would decide when she should visit the stallion, often adopting a programme such as mating on the third day after the first signs were seen and then again every other day until the mare went out of heat. In many cases, especially where the individual mare was experiencing a typical oestrous period of about five days, this rule-of-thumb approach was successful. However, when the individual was undergoing a prolonged oestrus, extending over many days or weeks, the mating schedule could not be completed without involving an excess of services. In many instances the studgroom might decide to limit the number of services to two or three, and thus the last service in the heat period might well have been at too great a time interval prior to ovulation for conception to occur. On the other hand, studgrooms developed an intuitive skill whereby they could interpret signs displayed by mares, and also interpret the way mares behaved in the paddock and in the stable. The studgroom thus developed the art of interpreting the mare's sexual signs by visual means.

This art has not been lost but it has diminished owing to the intrusion of the veterinarian into the field of interpreting a mare's sexual state by means of the rectal and vaginal examinations. These examinations became increasingly popular after the Second World War, largely following the work of John Hammond and Fred Day in Cambridge. They demonstrated the possibilities of confirming the oestrous state by observing the cervix through a speculum. More important, by palpating the ovaries for the presence of follicles they could determine the optimal moment for coitus and, further, cause ovulation at a predetermined time during the oestrous period, by the injection of hormones. The practical potential of these simple examinations was considerable, especially as at the time it was hoped the technique could be used to eliminate the problem of twinning. If at the rectal examination two follicles were palpated it was thought that, by various means, the incidence of twinning in Thoroughbreds could be eliminated or drastically reduced. Unfortunately these hopes have failed to materialise, for reasons discussed elsewhere.

However, the consequence of these aspirations was that owners and managers of studfarms became increasingly dependent on the veterinary examinations to determine the sexual state of the mare and for advice on the optimal time of mating. These methods caused the number of services by the stallion to be reduced, thereby saving the stallion and making significant economies in the man-hours devoted to taking mares to and from the covering barn. Further, by reducing the number of services necessary in each stud season, it became possible to increase the number of mares booked to the stallion. This practical advantage of the rectal examination was not fully utilised for many years because of the opposition of breeders to any increase in the traditional forty mares per stallion. This number was underwritten by legal arrangements for stallion syndication, wherein the ownership of the stallion was sold in terms of forty shares per horse.

In the 1950s, the habit of examining all Thoroughbred mares before service became so firmly established in the Newmarket area that it was not considered proper to have a mare mated unless she had been previously examined by a veterinarian. Part of this examination was to declare whether or not the mare was free from infectious disease likely to affect the stallion; mares were described as 'clean' or 'dirty' in this respect. Managers of stallions became accustomed to accepting mares only on condition that they had been 'seen' by the vet, and declared fit and ready for covering. In this way the vet became an integral part of stud management protocol; and the word of the vet took precedence over the opinion of the studgroom when it came to a decision as to whether or not a mare should be mated on any given occasion. The authority of the vet, in this aspect of stud management, became increasingly apparent in the 1960s, spreading from Newmarket to other breeding centres in England and Ireland. Indeed, the routine gynaecological examination of mares has become a common practice in most of the main Thoroughbred breeding centres of the world; the venereal epidemics of the early 1960s and the contagious equine metritis epidemics of 1977 added considerable momentum to the takeover by veterinarians of the managerial responsibility for mating programmes.

Venereal Infection: The Porchester Report

In the early 1960s the incidence of *Klebsiella* infection in mares appeared to be increasing. At this time, clinicians in the UK relied upon a visual inspection of the mare's genital tract to determine whether or not the mare was infected; little, if any, bacteriology was carried out. In Ireland, on the other hand, the collection of material from the cervix (cervical swabs) for bacteriological examination at the time of the routine inspections of the mare was widely practised. However, there was in both countries a difference of professional opinion as to the need for or efficacy of swabbing the cervix. This controversy culminated in the Thoroughbred Breeders' Association in England setting up a committee under the chairmanship of Lord Porchester. The committee reported in 1964 and 1966. The first report found that of 3,628 mares, nineteen per cent gave positive swabs at some time during the breeding season; and of these eight per cent were due to *Klebsiella pneumoniae*. The committee concluded thus:

> No evidence has come from the survey to substantiate the belief that infection with *Klebsiella pneumoniae* constitutes a specific venereal infection in the horse. There are firm indications however that *Klebsiella* may be passed on mechanically by a stallion from one mare to another mare. This organism can also be contracted by mares in other ways unrelated to service. No evidence is available to show that *Klebsiella* infection is an incurable disease, though its control may be difficult or prolonged when established in a mare.

On the subject of swabbing, the report provided the following recommendation:

> There is every justification for swabbing and culturing mares with uterine discharges, or those which are clinically abnormal, preferably early in the breeding season. The consideration of the 1964 survey

clearly indicates that it is not desirable to make any uniform, hard and fast recommendation regarding the swabbing of mares on all studs, since conditions vary greatly. Owners should be guided by their veterinary advisers, both for their mares at home and when visiting public studs, and should co-operate with the policy of the stud. It must be pointed out that swabbing is not an end in itself. The swab should identify the infection, and it is the prevention or the appropriate treatment and the successful elimination of infection which are important.

By 1966 the committee had developed a more conclusive attitude and recommended as follows:

After careful consideration of the facts disclosed by the 1964 survey and after full discussion the committee is unanimous in strongly recommending that:

1 *Barren, barren-maiden and maiden mares. All* barren, barren-maiden and maiden mares should be subject to bacteriological examination by cervical swabbing at least once before first service and subsequently at the discretion of the veterinary surgeon. Only those giving no growth of pathogenic organisms should be covered.

2 *Foaling mares.* Since some foaling mares have been found to be infected, the committee urge that foaling mares should also be swabbed when veterinary opinion considers it necessary.

On no account should any mare be covered until the veterinary surgeon is satisfied that she is clean.

In view of the evidence from two serious outbreaks of *Klebsiella* infection in 1966, where the spread of infection was by the stallion, it is in the interests of all mare owners to agree to the carrying out of the above recommendations.

The background to this change in attitude towards swabbing was the knowledge that 'two serious outbreaks of *Klebsiella*' had appeared on two studfarms in the Newmarket area during 1966. Thus, those who had pressed strongly for routine swabbing were vindicated, not so much by the overall survey of more than 3,000 mares but by a fortuitous epidemic that occurred on studfarms of high repute, and this was considered to be more significant than the eight per cent incidence of *Klebsiella* among infections recorded in the survey.

Contagious Equine Metritis

In 1977 another venereal epidemic appeared in the UK and Ireland, this time caused by an organism not hitherto identified and subsequently named *Haemophilus equigenitalis*. This infection appeared to be more contagious than any other previously encountered in horses; over fifty per cent of the mares at the English National Stud visiting the six stallions resident in 1977 were affected. Some seventeen studs in the Newmarket area and others in Ireland also reported having the disease on their premises. This extraordinary situation led to unprecedented commitments to veterinary control, based on the collecting of material from the genital tract of mares on several occasions before they would be accepted as 'clean' and fit for service. Codes of practice were developed and implemented by voluntary agreement, but backed by sanctions by which owners of mares who would not conform could not obtain service from any of the leading stallions.

In the late 1970s veterinary control of mating programmes among Thoroughbred mares thus became virtually complete. Fortunately, a close liaison between veterinarians and stud management has developed, and the system has worked efficiently, although the net effect is that few, if any, mares are mated in an oestrous period without the vet having first examined the mare. Further, mares are now tested before, and on, arrival at the studfarm during the breeding season. The Code of Practice is produced by the Horserace Betting Levy Board for the control of contagious equine reproductive diseases and can be obtained from the Thoroughbred Breeders' Association, Stanstead House, Newmarket, Suffolk CB8 9AA.

DIAGNOSIS AND TREATMENT OF INFERTILITY AND OTHER DISEASES

Palpating the ovaries for the size of the follicle and determining the sexual state of the mare are also the basis of diagnosing infertility, and in this respect the examinations are of a purely veterinary nature. The veterinary role may thus be defined as that in which the vet is dealing with matters of disease and pathology. This includes: (1) the diagnosis of the causes of infertility and subfertility in the mare and the stallion (see pages 245 and 263); (2) advising on, and implementing, preventive programmes against parasites (see page 32), epidemic diseases (see page 279) such as EIA, viral abortion, CEM and *Klebsiella* infections, strangles or against endemic diseases such as lockjaw; (3) dealing with the innumerable conditions and diseases which affect horses of all ages and types, such as injuries and colic.

VETERINARY RESPONSIBILITIES

The reader might be interested to consider the viewpoint of the veterinary clinician who has the responsibility of providing professional services to a studfarm. Perhaps the most onerous task is that of being available for emergency services at any time of the day or night, 365 days of the year. This requires a staff of at least two vets. Next in importance and resource-demanding effect are the technological possibilities made available by modern science. In the 'old days' vets could respond to the challenge of equine disease, infertility and other conditions with only the limited means available. Nowadays there is an ever-widening range of tests and therapies by which diagnosis and treatment may be attempted. The limits are more often defined by economics rather than by scientific criteria, and this applies equally to the client's as to the vet's resources. In fact, the ability of the veterinary profession to leave no stone unturned in any particular case may be directly related to the client's willingness to meet the cost of the required technology. For example, a mare who suffers from a twisted gut must receive immediate attention and can only be saved by a major abdominal operation, under general anaesthesia, followed by maximum effort to combat shock, including the administration of many litres of intravenous fluid. Both the diagnosis and therapy are based on laboratory analysis of blood for electrolyte and other content, employing skilled technicians using highly sophisticated instruments. Thus the best, most prompt and comprehensive veterinary service can only be achieved by teamwork and by incurring substantial financial outlay. There is inevitably some conflict between the possible (in terms of diagnostic and therapeutic effort) and the practical (in terms of what owners can afford or are willing to pay). The value of the animal is, of course, another important factor in the equation.

9
INFERTILITY

The subject of reduced fertility is for horse breeders one of obvious interest and significance, for it can result in commercial failure. Mares by their nature breed but once a year, and if for any reason a mare is barren one year, there is an inevitable two-year interval between times when she does produce. The loss in these cases includes the annual cost of keep, the stud fee of the stallion, and the depreciation in value of the mare, who at best has a productive life of about fifteen years; thus a year of barrenness represents between one-fifteenth and one-tenth of possible output. Subfertility is a problem affecting both mare and stallion. The problem of one may add to that of the other.

Definitions

The terms *infertility, subfertility* and *sterility* require some definition for the purpose of our discussion. Infertility and subfertility are interchangeable terms and both imply that the individual has a below-average ability to conceive and carry a foal to full term. Any definition must be somewhat imprecise because of the many variables affecting the results of matings, from the point of view of the mare and of the stallion. A mare may be described as infertile during one breeding season but fertile in another. The same qualifying description may be applied to one heat period compared with another, eg the foaling heat is said to be a less fertile heat than those that occur subsequently.

In practice, the term subfertility is used to describe mares which are 'difficult' to get in foal or which breed infrequently. If we study the performance of a random number of individuals, we find the following patterns of breeding can be identified as occurring in the population:
(a) Individuals that give birth to a foal annually throughout a relatively long breeding life.
(b) Individuals that breed successfully until middle age, after which they are less productive.
(c) Every-other-year breeders.
(d) Individuals that go through alternating periods of successful foaling and failing to conceive or aborting.

(e) Mares that abort habitually at some stage of pregnancy or conceive twins.

A subfertile or infertile mare may be defined for practical purposes as one that produces a live foal in fewer than four years out of five. From a veterinary viewpoint, the definition is that of a mare with a known problem reducing the chances of conception and a successful pregnancy.

A sterile individual is one where there is a condition which prevents the mare from breeding at any time. Two examples of such a condition are congenital abnormalities, and senility.

In the stallion, subfertility implies a less-than-average ability to obtain conception in a group of thirty to forty mares. The definition is quite arbitrary because the fertility of the mares is a variable factor that may affect the result in any particular case. Further, management, seasonal and other factors may affect the outcome. Insurance underwriters usually insure on the basis of assuming that an average horse will obtain more than sixty-per-cent conception in a group of twenty to forty mares during the breeding season. The sixty-per-cent level is chosen because the majority of stallions obtain a somewhat higher ratio than this.

A veterinary definition has to take account of abnormal sexual behaviour (eg failure to mount the mare or to ejaculate) and deficiencies in semen quality. Veterinarians recognise subfertile individuals by the pattern of conceptual failure among fertile mares during the breeding season.

A Mare's Infertility

We should not think of infertility solely in terms of a mare failing to conceive. The fertilised egg is vulnerable during the whole period of its development from single-cell size to the mature foetus at foaling time. We shall find many causes of abortion to discuss in chapter 10, each of which may prevent an individual producing a foal in any given year. Reproduction is a continuing process, one stage depending for its normality on that preceding it. Infertility in the mare is not, therefore, a simple question of getting in foal or being barren. This simplistic approach ignores the full extent of the problem.

A Stallion's Infertility

The fertility of the stallion is based on the conceptions obtained in a group of mares, and the basis on which these results are obtained may vary with the endpoint of the particular pregnancy test selected for the purpose. It will also be affected by the breeding status of the mares and the number that are mated during the breeding season. A high proportion of poten- tially infertile or 'difficult' mares will tend to depress the percentage fertility rate, however this is assessed. And the greater the number in the group, over about forty, the lower the expected fertility rate. Numbers are important in practice because the more mares that are booked to a horse, the greater will be the number of subfertile mares among them; and the greater number of services for more mares will result in decreased numbers of sperm per ejaculate.

The most reliable, and probably the fairest, means of estimating the fertility of a stallion is to assess his performance on maiden mares. Ninety per cent of maiden mares conceive in their first breeding season, and a fertile horse would be expected to obtain this percentage. The minimum number that could usefully be used as a test would be eight individuals, although some indication could be gained with less.

Influences in the Equation of Fertility

However we define subfertility, the outcome in any given situation is a balance of influences involving the basic capacity of the individual to breed, the environment, the standard of management, and an interaction between the fertility of the mare and the stallion. Management plays a decisive role because, apart from the health of the mare and stallion, all other factors are more or less under managerial control.

INFERTILITY IN THE MARE

Infertility may have managerial, physiological or pathological causes, or a combination of these. The condition may be temporary (ie for a breeding season only), for a long term (lasting for several breeding seasons), or permanent (sterility).

Management

Managerial reasons for mares not conceiving include:

(a) Failing to present the mare to the stallion at the optimal time prior to ovulation due to (i) silent heat, (ii) mating too early in oestrus, (iii) mating after ovulation has occurred.

(b) Failure to recognise that the stallion has not ejaculated.

(c) Failure to take necessary preparatory steps, such as asking for or accepting veterinary advice. For example, a mare takes air into the genital tract but no measures may be taken by management to diagnose or to remedy this by a Caslick operation on the vulva. Similarly, it is often not appreciated by management that the diagnosis of problems *and* their correction is best performed in the months outside the breeding season in those cases where there are pathological problems of endometritis (inflammation of the uterine lining). We shall see later how important the coital challenge is to mares in a susceptible condition and the extent to which management may mitigate or exaggerate the consequences for these individuals.

(d) Mistakes made by veterinarians during the rectal examination may contribute to a mare being 'missed' by, say, miscalculating the expected moment of ovulation, so that a mare is not mated until after the event and goes out of heat before she can be mated.

The importance of errors in management is exaggerated by the arbitrarily restricted limits of the breeding season, against the background of the physiological and pathological influences considered below. Conception can only occur if ovulation takes place and the sperm are in the Fallopian tube within a relatively short period, 24 to 48 hours. The number of ovulations per breeding season represents the number of chances for conception to occur; and this number is limited by the arbitrarily defined limits of the season.

Physiological Factors

If a mare were to ovulate every twenty days from the beginning to the end of the stud season (150 days) there would be seven eggs shed during this period. In fact, the following influences diminish these opportunities:

(a) The oestrous cycle, in many individuals, is longer than twenty days especially in the first two months of the breeding season.

(b) Ovulation does not occur in every oestrus, again especially during the first two months of the breeding season.

(c) Some mares suffer prolonged dioestrus that may cause them to be refractory for six to twelve weeks or more unless treated. This prolonged dioestrus may take place after mating and treatment may, therefore, not be advisable until, at least, the earliest time for an accurate pregnancy diagnosis to be made, usually about forty days.

(d) Ovulation may occur without oestrous signs, or immediately prior to a mare coming into oestrus. This asynchrony is physiological, ie it is normal, and in the natural situation the mare might be mated by the stallion prior to the event. However, under the conditions of breeding in hand the mare shows signs too late for a planned mating and the individual is thereby not served until too late.

Taking these physiological influences into account, it may be estimated that the number of ovulations occurring during the 150 days of the stud season may in many instances be reduced to only three or four. The importance of utilising to the full each of these ovulations is abundantly clear if we consider the possible adverse effects of management and pathology, as applied to the mare and the stallion. A missed heat, failure of a horse to ejaculate or the ejaculation of a subfertile semen sample reduces the chances of conception. In each heat period where this occurs the available chances during the breeding season are thus reduced by a third or a quarter. Pathological states further diminish the chances of conception.

Pathological Influences

Disease of the genital organs may reduce or entirely eliminate the chances of conception occurring during one particular oestrus, for several periods or indefinitely. Thus, disease may result in subfertility of a temporary or seasonal nature or impose a long-term problem. Let us consider a number of examples of how disease may reduce the chances of a mare conceiving or carrying a foal to full term.

Many of the conditions of pathological infertility involve the genital tract and, in particular, the uterine lining (endometrium). In the first example we may suppose that a mare has become infected with a microbe such as *Klebsiella* or the contagious equine metritis organism. The microbe is passed into the uterus by the stallion at the time of coitus. It invades the uterine lining, causing the lining to become inflamed. The presence of the microbe destroys the fertilised egg as it enters the uterus, so that the mare does not conceive in that heat period. The inflammation increases in severity as the microbe breeds on the surface and within the lining itself. A discharge of pus cells and mucus, secreted by the glands in the uterine lining, escapes from the uterus through the cervix into the vagina. The discharge may be seen coating the vulval lips and the perineum, tail and inside of thighs and hocks, depending on the quantity of exudate produced by the mucous membrane lining of the uterus. The external signs may, therefore, be more or less obvious depending on the uterine response. But at the next oestrous period, if the mare is mated without treatment, the microbial growth is stimulated by the presence of semen in the uterus and this results in an increased death rate of the sperm, and thereby a diminished chance of the egg becoming fertilised. Further, should the egg be fertilised, the presence of the microbe and the inflamed surface of the uterus prevents survival once the developing foetus arrives in the uterine environment.

We can appreciate, therefore, that uterine infection, if left untreated, may cause infertility not only during one but in several oestrous periods. In fact, the longer the infection is present the more severe are the inflammatory changes in the uterine lining. These changes increase the chances that conception will not occur or of early embryonic death. Even if the infection is cleared by treatment, the chronic uterine changes, which include fibrosis and glandular degeneration, may persist and be the cause of subfertility, that is, a reduced chance of conception and of foetal survival.

Pneumovagina is a condition in which air is abnormally drawn into the genital tract. The air carries small particles of dust and microbes which cause irritation and infection of the uterus. So long as the defect in the arrangement between the vulva, perineum, vagina and pelvis is present the individual will remain infertile. When the vulva is sutured in a Caslick operation the integrity of the valve arrangement is restored, and air is no longer drawn into the vagina. The infection in the uterus subsides, and the chances of conception increase dramatically. However, if the condition has been present for some lengthy period, damage to the uterine lining may have become severe and thus be responsible for continuing subfertility.

These examples serve to show how uterine damage may reduce the chances of conception, even though ovulation has occurred. There are instances where the ovaries become affected by tumours and prevent the shedding of eggs. There are also conditions of hormonal dysfunction, but this is an area which is poorly defined and in many cases the dysfunction is physiological and related to the season or the environment (see above) rather than being pathological. The exception is the presence of infection in the uterus which disturbs the prostaglandin mechanism and thus causes mares to exhibit very short oestrous cycles or, alternatively, not to come into oestrus at all, depending on the nature of the damage to the uterine surface. The term 'hormonal dysfunction' or 'imbalance' is one often used by clinicians and studmen but it has little precise meaning unless carefully defined and the causes accurately diagnosed (see Diagnosis, page 255).

If we return to the idea that the average mare ovulates only three or four times during each breeding season, and add to this the risks of physiological aberrations and pathological disorders, we can appreciate the problem of subfertility. Of course, we need not suppose that all mares are restricted to this low number of ovulations; nor that one ovulation is necessarily insufficient for a mare to be got in foal.

Diseases of the Genital Organs

Pneumovagina The normal arrangement of the perineum, vulva, vagina and cervix is shown in fig 4 (see page 67). It will be noted that the majority of the vulval length is below the pelvic brim, and that the tube formed by the vagina is closed by 'valves' at the vulva, the vestibule and the cervix. Air is thus prevented from entering the genital tract although during oestrus, when the vulva is relaxed and lengthened, some air may enter the last part of the tract. It is prevented from entering further by the vestibular valve. In addition, the protective mechanism against microbes includes the secretion of mucus containing antimicrobial substances that deal with microbes that happen to reach the cervix.

The protective seals may be breached by one or a combination of the following situations:

(a) The conformation of the perineum relative to the pelvic brim may be poor in that the majority of the length of the vulva is above, and not below, the brim. This conformation is based on the structure of the area and is, to a degree, inherited. This relative elevation of the entry to the tract has two effects; it tends to destroy the integrity of the vestibular seal and it allows the vulva and perineum to decline in a forward direction, so that they form an angle with the perpendicular, in some cases coming to lie almost in the horizontal plane. Dust and microbe-laden air enters the tract through the vulval and vestibular seals, much of the material coming from the faeces as it passes over the vulva.

Normally, the walls of the genital tract are closely apposed to each other. However, the surrounding organs are arranged in such a manner

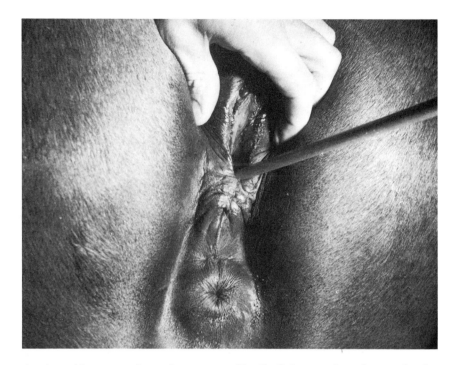

A vulva with poor conformation corrected by Caslick operation; the metal rod is resting on the floor of the pelvis

that there is some degree of suction exerted on the vagina and uterus, in which a potential vacuum exists. If the tract is exposed to atmospheric pressure through open vestibular and vulval seals, air will then be sucked into the vagina and, if the cervix is open, into the uterus.

(b) During oestrus, the walls of the vagina become moist, and the tissues supporting them, including the vulva and cervix, are relaxed. This makes it easier for the walls to separate at the time of coitus, but also inevitably allows the vestibular and vulval seals to open. Thus oestrus is a period during which the genital tract is particularly at risk to the entry of air and infection. The risk is greatest where there is poor conformation. The entry of the male organ, at the time of coitus, increases the problem because it carries a large number of microbes deep into the tract.

(c) At foaling, the foetus passes through the birth canal composed of the cervix, vagina and vulva. This stretches the genital tract at a time when the uterus itself is stretched to a maximum capacity. Although the walls of the birth canal and of the uterus contract quite quickly after foaling, they are still relaxed for many hours and only completely return to the barren state some days, or even weeks, after foaling. Most readers who have attended a foaling will have heard the sound of air rushing into the genital tract. This occurs most often as the mare stands up after delivering her foal, with the placental membranes (afterbirth) partly in the uterus and partly hanging down behind. In this situation, the vulval lips are parted by the extruding membranes, the genital tract is relaxed, and the vulval, vestibular and cervical seals are open. In passing we should note the harmful effect of a dusty atmosphere, caused by poor-quality straw bedding, especially if shaken vigorously. The dust carries fungal spores and microbes which, if sucked into the uterus at this time, are responsible for fungal infection, causing infertility or abortion in the next pregnancy.

(d) The vulval lips may be slack and gaping so that the integrity of this seal is completely or partially lost. In these cases air penetrates into the tract only during oestrus when the vestibular seal is relaxed.

(e) The ligaments and tissues supporting the uterus and other parts of the genital tract may become slack, especially with age. The suction on the walls is then exaggerated, and there is an increased risk of air being drawn into the tract if there is the slightest weakness in the protective seals. In some of these cases the cervix can be viewed through a speculum positioned below the brim of the pelvis. The operator has to direct the

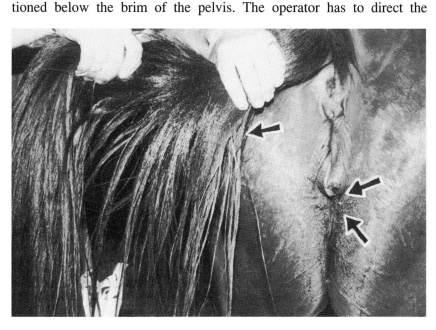

Mucus and 'pus' cells (leucocytes) from the uterus are seen as sticky or dried discharge (arrowed) on the tail, vulva and inside of the thighs

speculum at an angle downwards, towards the cervix, rather than in a horizontal direction or slightly upwards. Urine or fluid is often found pooled on the floor of the vagina.

If a speculum or small tube is inserted into the vagina, air will normally pass into the vagina, sucked in by the potential vacuum in the genital tract. However, if air is already present because the protective seals are not properly functioning, the introduction of a speculum will not have this effect.

Inflammation of the uterine lining The lining of the uterus consists of an outer epithelial lining under which is a submucosal layer of glands and connective tissue. Microbes and other factors may damage the lining causing inflammation (endometritis). The term 'acute endometritis' denotes an inflammatory response to damage, the feature of which is the presence of pus cells. These may accumulate in the submucosal layer beneath the epithelium, through which they pass together with mucus secreted by the glands. We can see the discharge of pus cells (leucocytes) and mucus as it leaves the genital tract of 'dirty' mares.

The term 'chronic endometritis' is used to describe the presence of other types of inflammatory cells. For example, plasma cells and lymphocytes are special white blood cells concerned with the body's immune response. The larger white cells, known as mononuclear cells, are concerned with 'mopping up' debris and microbes after the initial damage and are commonly present in chronic endometritis. Chronic endometritis includes changes which are long-term and less easily treated than the acute condition. The reader should appreciate that the terms acute and chronic are those applied by the pathologist and do not necessarily indicate the severity of the condition.

Hormonal imbalance This term is often used by vets and horse owners, but it has no precise meaning and probably little practical relevance. Most of the failures in the oestrous cycle, especially those involving oestrus and ovulation, are more affected by season and the environment than by pathological causes. There may be cases of primary hormonal dysfunction but we cannot diagnose these at present because of limits to our diagnostic techniques and understanding. Hormonal dysfunction is a feature in only two recognised conditions:

(a) The granulosa cell tumour of the ovary interferes with hormonal function, and the opposite ovary, in consequence, becomes small and inactive. The presence of the tumour prevents ovulation and although some cases show oestrus, many affected mares do not.

(b) Congenital conditions of the ovaries, causing them to be minute, or even absent, are associated with complete ovarian inactivity. The condition is due to defects in the sex chromosomes. The female normally has two sex chromosomes shaped roughly as an X (ie XX) and the male has one X and one Y chromosome (ie XY). In these cases, the mare's cells may contain only one sex chromosome, described as XO, an extra chromosome, XXX, or a variety of other abnormal combinations.

Senility Mares become sexually senile at different ages, but most commonly at age eighteen onwards. The ovaries become small and inactive, and the uterine lining loses its glands.

Anatomical abnormalities Apart from malformation of the perineum, anatomical abnormalities play a very small part in infertility. Imperforate hymen is a condition of maiden mares where the pillars surrounding the vestibular seal form a complete membrane. This condition is uncommon, and the membrane may be opened by means of a small operation. More substantial abnormalities such as the congenital absence of a cervix are, of course, much more serious and result in complete sterility. Fortunately, these abnormalities are very rarely encountered.

Infections Microbial infections are commonly associated with infertility. But it is not always easy to determine whether, in a given case, the recovery of a microbe from the uterus indicates a symptom or a cause of the condition. Table E lists the microbes that may be grown from material collected from the genital tract. The list is divided into primary and secondary pathogens, ie those which damage the lining of the tract with (primary) or without (secondary) the presence of a predisposing factor. Further, the nature of the organism may change, and in some circumstances a secondary pathogen may behave like a primary pathogen and vice versa.

Predisposing causes include pneumovagina, the aftermath of foaling and gestation, and varying states of the oestrous cycle. For example, the uterus is more susceptible to infection during dioestrus and anoestrus than it is in oestrus. Old age, trauma or bruising by foaling, medical procedures involving distention of the vagina and cervix, abortion, malnutrition and debility or toxaemia due to generalised conditions or infections are other factors predisposing to uterine infection.

TABLE E Some primary and secondary microbes affecting the mare's genital tract

Primary Pathogens

Klebsiella aerogenes	Types 1 and 5
Haemophilus equigenitalis	Contagious equine metritis (CEM)
Pseudomonas aeruginosa	Some strains

Secondary Pathogens

Streptococcus zooepidemicus	
Escherichia coli	
Staphylococcus aureus	
Klebsiella aerogenes	Types 7 and others
Pseudomonas aeruginosa	Some strains
Fungi	

Contaminants

Staphylococcus albus

The susceptible mare The term 'susceptible', in this context, means susceptible to infection and inflammation of the uterine lining. The susceptible mare is recognised in practice as one that frequently becomes infected although not necessarily with the same microbe. We may isolate an *E. coli* on one occasion and a *Streptococcus* on another. The usual pattern is for the mare to appear 'clean' when in dioestrus but in oestrus show signs of a discharge or other evidence of infection. The most typical pattern is a vaginal discharge one or two days after coitus, showing that the mare is susceptible to the coital challenge (see below). The defence mechanisms against infection, at the uterine surface, involve special proteins IgA and IgG. These proteins contain the antibodies that kill the microbes by direct and indirect means. Antibodies may cause microbes to clump together or they render them more accessible to scavenger cells, such as pus cells, which ingest and kill them. The uterine defences depend on these immune bodies and the scavenger cells. The latter are promoted by oestrogen, and are therefore more effective during oestrus than at other times of the cycle. However, the signs of infection are more obvious during oestrus, when the mucus lining the genital tract becomes copious and freely moving. Further, the presence of polymorphs in increased numbers gives the secretion a thick or purulent appearance.

We do not know why the immune mechanisms fail in certain individuals and on particular occasions. Gestation and foaling may play a major role in the problem, because the uterine defences are probably weak in the days after foaling. They normally become restored, sometimes by the foaling heat, and in most cases by the subsequent heat period. It is obvious in practice, however, that restoring the normal defence mechanisms takes longer, or even fails to be entirely restored, in some individuals. Age is another factor which seems to play a significant part in the susceptible mare syndrome, particularly affecting individuals over about age eleven years.

Coital challenge The penis of the stallion harbours large numbers of microbes, most of which are innocuous and incapable of causing infection in the mare. However, this population of microbes may be altered by circumstances. Coitus in a mare that is, herself, carrying primary pathogenic microbes, or a particularly large infection of secondary microbes, may significantly change the microbial flora. Treatment of the penis with disinfectants, antiseptics or antibiotics may remove, at a stroke, most of the microbes on the surface. But the more that the organ is cleansed the greater the risk that it will be repopulated by potentially harmful microbes such as *Klebsiella* and *Pseudomonas*.

It is inevitable that many microbes are inseminated into the posterior part of the mare's genital tract at service. Some of these microbes are innocuous and others may stimulate the defence mechanisms of the uterus and cervix which readily overcome them. However, the coital challenge is significant in a susceptible mare whose defences cannot deal with the microbes. These susceptible mares should be inseminated artificially, or measures taken to support the defence mechanisms by, say, introducing antibiotics and other protective substances before, at the time of and/or after coitus.

Diagnosis

The diagnosis of the cause of infertility is perhaps the most ex cting problem facing the veterinarian in routine equine studfarm practice. In most cases of disease, the diagnosis depends largely on current symptoms and the results of laboratory tests. Infertility, on the other hand, is a rather more nebulous and wide-ranging condition. In each case we have to consider the past, present and future managerial and environmental influences, and the possible pathology. An exhaustive inquiry into the history of the mare may be necessary to discover the reasons for her not getting in foal; and observation at intervals may be required before we can come to any conclusion. To a large extent the diagnosis depends on the outcome of the case and this must be judged in terms of conception and delivery of a live foal. The following are the steps which may be taken to achieve a diagnosis.

1 History of mare (a) Results in previous breeding seasons should be examined to determine the pattern of breeding. For example, some mares are difficult to get in foal or may be infertile during the time they are suckling a foal. These every-other-year breeders often present no problem when they start the breeding season in a barren state.

(b) A knowledge of sexual behaviour in previous breeding seasons may be helpful, although mares do not necessarily exhibit the same patterns each year. For example, a mare may be said to show little signs of oestrus before ovulating, but in the next stud season she may well exhibit strong signs of oestrus for many days before developing an ovarian follicle and ovulating. However, some individual behaviour is consistent from year to year, and such knowledge may prove invaluable to stud management.

(c) The veterinary history is perhaps of most importance. Signs of vaginal and uterine discharge, the bacteriology of the genital tract, and the size and activity of the ovaries in the past may provide important clues to problems that occur during a current breeding season.

(d) The number of foals carried, years barren or pregnancies aborted in the past provide a useful guide to the likely outcome of current matings. A barren mare may have had a history of barrenness only in the immediate past breeding season. She should thus be still regarded as a mare with high potential unless she were a maiden mare. Maiden mares have a ninety-per-cent record of conception, and to have been barren as a maiden therefore suggests some special problem. The next step in this case, as with an older barren mare, would be to assess the previous year's performance for evidence of whether or not the mare is likely to have some particular problem. If the mare had not been mated on more than one occasion, there would be reason to suppose that the cause of her not getting in foal was more physiological than pathological. By contrast, a mare that had been mated on four or five occasions could be presumed to have some definite problem – unless, of course, the stallion could be shown to be subfertile. Mares who have been barren and/or aborted in more than one in three past breeding seasons, and those that have aborted single or twin foals or have experienced an abnormal pregnancy or foaling in the past twelve months, are liable to prove difficult to get in foal and may require special measures of diagnosis and therapy.

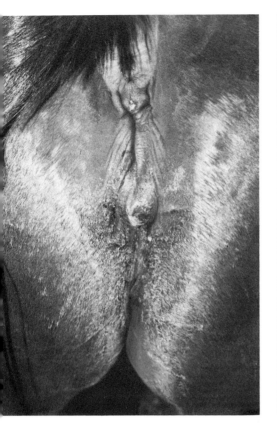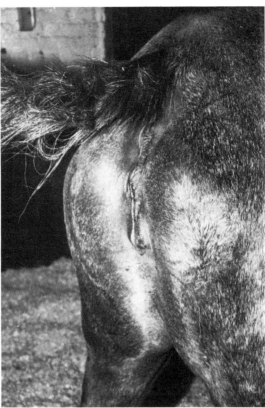

Studmen should familiarise themselves with the signs of a vaginal discharge, shown here (left) *coating hairs around vulva and* (right) *matting hair of tail*

(e) An analysis of breeding returns from the Thoroughbred Stud Book shows a significant fall in breeding potential with age. In assessing the chances of fertility in a given year, the age of the individual must therefore be taken into account. Similarly there is a tendency for mares of exceptional racing ability to show subaverage fertility.

The breeding history of a mare can, therefore, be used to identify individuals who are likely to have a problem. It should be stressed that these mares should be examined and, if necessary, treated between the breeding seasons and not, as is so often the case, at the start of, or during, a breeding season.

2 *Rectal and vaginal examinations* form the basis on which the diagnosis of subfertility is made. The interpretation of findings for diagnosis of

subfertility, rather than for managerial convenience, may require a series of examinations to assess changing patterns and to establish if abnormal structures in the ovaries or uterus are a continuing or temporary state. Of course, this information has to be interpreted against the background of the time of year, body condition, hair coat length, nutrition and other influences. The same necessity for serial inspections may apply to vaginal examinations.

The perineum, vulva and conformation of the genital tract is examined for evidence of pneumovagina, discharge òr inflammation. Stud personnel should make themselves familiar with the external signs of genital tract problems, such as vaginal discharges. A discharge, in itself, may not be significant, but its presence should be noted and reported to the vet, particularly as it might be caused by one of the primary pathogenic microbes responsible for venereal disease. All discharges from the vagina should be regarded as abnormal. However, in some cases the sediment on the inside of the thighs, due to deposits of salts in the urine, may be mistaken for a vaginal discharge.

3 Bacteriology of the genital tract (swabbing) The cervical swab is now a well-recognised aid to veterinary diagnosis. In fact, the verb 'to swab' has become part of studfarm language. Swabbing is a technique for collecting material from any site in the body. The swab consists of cotton wool or other absorbent material attached to a rod of wire or stick of wood and maintained in a sterile condition. Material for examination can thereby be collected, whether from the nasal passages, an abscess or the genital tract, and transported to the laboratory for culture. The questions we must ask ourselves are: (a) What is the best site from which to collect the material in the relatively lengthy genital tract of the mare? (b) What is the significance of microbes grown from the material collected from that site?

Site – the most convenient site for the collection of material is undoubtedly the posterior end of the tract. The cervix is readily accessible through a speculum and it is possible to pass a swab into the cervix without difficulty. The cervical swab has therefore been the one on which vets have mostly relied for evidence of infection.

In recent years, attention has been drawn to two other sites which may harbour microbes, the opening of the urethra on the floor of the vagina and the clitoris. Swabbing these two sites has now become routine practice. *Klebsiella,* in particular, may 'lurk' in the urethral orifice, and the CEM organism in the clitoris. The clitoris possesses a body, a cavity or fossa and a sinus in which smegma may accumulate. It is in this oily paste-like substance that microbes may be harboured, and it is therefore necessary to swab from the sinus or to squeeze the smegma to the surface so that it may be tested for the presence of CEM or other organisms.

Significance – the significance of microbes cultured from any part of the genital tract depends on their nature and identity. The primary pathogens are potentially harmful from whichever site they are recovered. These microbes, if transferred to another mare at coitus, are capable of damaging the uterine lining and of establishing themselves as a venereal disease. Recovery of these microbes from the cervix, urethra or clitoris must be followed by strict precautions and the withholding of coitus until such time as the microbe has been effectively eliminated.

The secondary pathogens may be transmitted by coitus from one mare to another. They do not cause disease in a healthy mare but may do so in one that is susceptible. These microbes are not regarded as significant if they are cultured from the urethra or clitoris because these sites normally harbour many types of bacteria. In contrast, the cervix and uterine lining are normally sterile; if microbes are recovered from either of these two sites, they should be regarded as being of potential significance for that individual. Microbes such as *Staphylococcus albus* are an exception because they cannot cause disease and are regarded merely as contaminants.

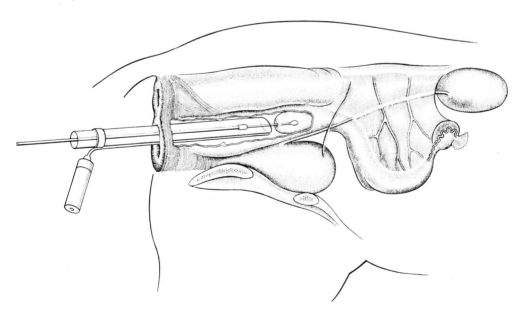

Fig 27 Cervical swabbing consists of passing a swab attached to a metal rod through the speculum into the cervix

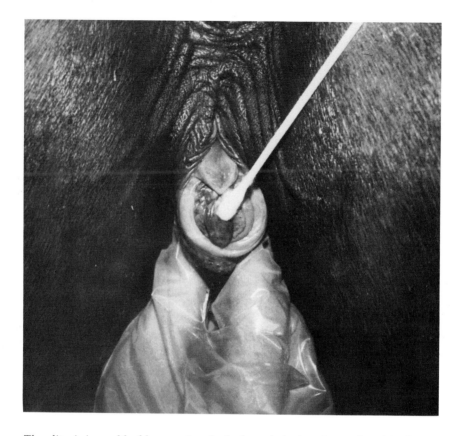

The clitoris is swabbed by everting the body and placing the swab end in the clitoral fossa; smegma may be squeezed from the sinuses or a finer swab passed directly into the sinus

The optimal time to take samples from the cervix is during the early part of oestrus, when it is relaxed and the secretions are fluid rather than sticky. The sample is then more representative of the whole of the genital tract than is the case with a mare in dioestrus. Samples from the clitoris may be taken at any time because this site forms a reservoir, reflecting past secretions that have flowed over its surface.

4 Cytology The study of cells obtained from smears of the uterine lining or cervix is a necessary and reliable means of assessing the significance of microbes grown from the cervix. We can assume that the uterine lining is inflamed if certain white blood cells (polymorphonuclear leucocytes) are present in the smear. These are not normally present, except perhaps for a

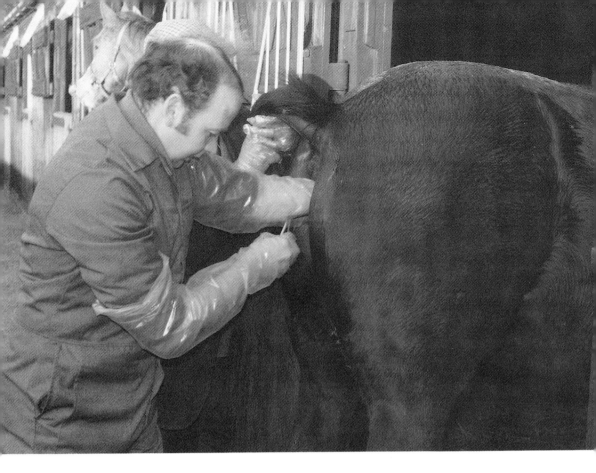

Obtaining a biopsy specimen from the uterus of a mare

day or two after a mare has been mated. Thus the test provides evidence which, taken in conjunction with the bacteriological findings, is helpful in diagnosis.

5 *Uterine biopsy* The techniques of uterine biopsy (fig 28) enable us to make a microscopic examination of the uterine lining by means of a simple and safe technique. The biopsy instrument has a small pair of jaws that can be operated by the handle outside the mare. A small portion of uterine lining is taken into the jaws and transferred to special fluid for preservation, and examined in the laboratory.

Here the expert can determine the degree and type of inflammation, thus providing the clinician with valuable information regarding treatment and the probable future outcome of the condition. This is not the place to describe details of uterine pathology, but in general terms the pathologist may report an acute inflammation (endometritis) denoting an active infection at the time of sampling, chronic conditions indicating long-term changes, degenerative glandular abnormalities, fibrous scarring, and senile or immature conditions.

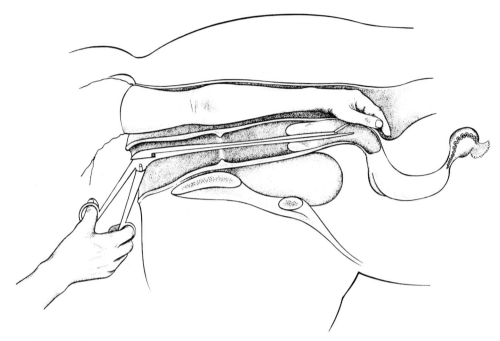

Fig 28 The technique of uterine biopsy: the operator's arm presses on the instrument which is inserted through the vagina and cervix into the uterus so that a small pinch of the wall of the uterus may be obtained by the instrument

6 Other aids to diagnosis Modern optical instruments can be used to enable vets to look into the uterus by passing a flexible fiberoptic endoscope through the cervix. Other similar instruments may be passed through the abdominal wall, giving direct viewing of the ovaries and outside of the uterus. Serology (the examination of blood for antibody levels) may be used to detect the body's response to the presence of microbes. Thus serology may be used to condemn a mare that is positive but not to clear a mare that is negative.

7 Measurement of hormone levels in the blood stream of mares during the oestrous cycle can contribute greatly to diagnosis. Progesterone may be measured to determine whether or not an active yellow body is present in the ovary at any given time. Measured at intervals over a period, progesterone levels enable us to determine whether or not a mare is 'cycling', when taken together with clinical findings in the case. In special circumstances the measurement of other sex hormones may help in the diagnosis of the cause of an individual's failure to conceive.

INFERTILITY IN THE STALLION

In chapter 4 we discussed the stallion's sexual functions in terms of libido (the desire and ability to mount the mare, achieve intromission and ejaculation) and the capacity to deliver semen of sufficient quality and quantity to fertilise the mare's egg. Subfertility is the other side of this coin, namely that of diminished ability to deliver semen and/or a reduced fertilising capacity.

Insufficient libido, failure to maintain an erection, inability to gain intromission or to ejaculate, may be due to psychological causes or to pain resulting from injury or disease. The sexual act is a sequence of behavioural patterns, a failure in any part of which will prevent the next and subsequent phases. The final phase, ejaculation, may be prevented by painful lesions in the back or on the penis, causing the horse to dismount before ejaculating. Management has an important responsibility to identify whether or not a horse has ejaculated on any particular occasion.

The Fertilising Capacity

The fertilising capacity of semen depends on the presence of sufficient numbers of normal live sperm. Any factor which reduces the likelihood of this number being achieved will tend to make the individual subfertile. The following conditions may, singly or in combination, be involved.

(a) A high proportion of dead sperm will, obviously, diminish the stallion's fertilising capacity by a corresponding degree.
(b) Sperm may lose the ability to live or 'swim' in the genital tract for sufficient time to allow them to travel to the Fallopian tube or to survive until the egg is available to be fertilised.
(c) The sperm may have structural defects that prevent them from living, moving or being able to fertilise the egg. These defects may be classified as primary, secondary or tertiary according to whether they arise at the site of sperm production, in transit or after ejaculation. Among primary defects is 'knobbing', which is a deformity of the acrosome covering the sperm head; protoplasmic droplets are secondary defects; and broken or bent tails are tertiary defects, often resulting from trauma at the time of collection into an artificial vagina.

The classic method of examining semen is to collect a sample through an artificial vagina and to analyse the sample in the laboratory for numbers of sperm (by concentration per millilitre and in total within the ejaculate), the percentage of live/dead and of normal/abnormal forms. Some estimation of motility is also made in percentage terms at hourly or half-hourly intervals to assess the longevity of sperm within the sample under standard conditions of temperature, etc. The methods used in the laboratory need not concern us here except that they include the use of various staining and microscopic techniques.

The interpretation of laboratory analyses of semen relating to expected fertility is limited by variations in content and quality according to the season of the year, the number of ejaculations per week and the method of collecting the sample. It may be necessary to obtain a series of samples under comparable conditions to evaluate a stallion that produces a sample of borderline features. Experience shows that under existing methods and knowledge, a forecast of a horse's fertility, based on laboratory results, cannot be 100 per cent accurate in any case. There are a number of exceptions in which, apparently, poor-quality semen is ejaculated by stallions with relatively high fertility rates; conversely, occasionally excellent samples are obtained from horses that are demonstrably infertile.

The status of the stallion has some bearing on the conception rates. A much sought-after stallion with a relatively costly nomination fee will inevitably attract mares that are selected, partly, on the basis of a good breeding performance. One would expect that forty mares visiting, say, a Triple Crown winner with a nomination fee of £15,000 would have better breeding potential than those sent to a horse whose nomination was £400 'no foal no fee'. Fashion may well determine the eventual fertility percentage in any given year.

The minimal number of live normal sperm in an ejaculate necessary to achieve conception in a mare of normal fertility is said to be in the region of 100 million. Let us consider two examples of how this figure may be achieved in practice. Stallion A ejaculates a semen sample of 50ml with a density of 200 million sperm per ml, ie 50 x 200 = 10,000 million per ejaculate. Of these, fifty per cent are normal. Stallion A therefore has an effective ejaculate of 5,000 million, ie fifty times the assumed normal minimal level. His ejaculates are likely to be fertile, even if the numbers of sperm are depleted by over-use. Stallion B, on the other hand, has a smaller ejaculate, less density but a higher proportion of normal live sperm. His figures are as follows: an ejaculate of 20ml containing 50 million sperm per ml, ie 1,000 million sperm per ejaculate. Twenty five per cent of these are normal, ie 250 million sperm per ejaculate. Thus stallion B's ejaculates contain 250 million normal live sperm. This is very close to the minimal level for fertility. The reader should understand that these examples are used to illustrate the variations that may occur between stallions and the number of sperm inseminated into the uterus. However, these variations may occur within the individual during a breeding season; the quality of a service may vary from day to day. If, in the examples above, the quality of the ejaculate is consistent and representative of the average delivered by stallions A and B, we may assume that stallion A will have a high fertility whereas stallion B may only get his mares in foal if the conditions are in his favour. This may require that stallion B be allowed only one service per day and mates with mares who are fertile and presented under conditions of management ensuring that mating occurs near to the time of ovulation.

Breeders recognise the fact that some stallions have semen of long-lasting qualities, while others require matings to occur within 12 to 24 hours of ovulation. There are recorded cases of a stallion having mated successfully with an individual at two, three or even seven days prior to ovulation. However, examples of the opposite situation are more common, ie stallions that do not appear to be able to achieve conception in their mares if mated beyond two days. In fact, this is so much the rule with Thoroughbreds that it is customary for mares to be cross-served (remated) at two-day intervals. No serious study has been made of this aspect of fertility, but it is reasonable to assume that stallions with longlasting semen are those with a high number of normal live sperm. The greater the number of inseminated sperm, the longer will minimal numbers be sustained in the face of the inevitable fall-off in numbers as sperm lose their motility and ability to fertilise. Of course, we cannot rule out the possibility that some stallions have better-quality semen and their sperm live longer due to an ability to survive in the mare's genital tract for longer periods. There is some evidence that the seminal plasma of the stallion may contain substances which kill the sperm after a certain period. This spermicidal (killing) action is thought to be necessary to avoid deterioration of the sperm due to ageing. Ageing may produce genetic defects in the fertilised egg, and thus some measure to reduce the risk of this biological quirk of nature may be necessary. It has been found in some species that these killing substances are overactive in some individuals, limiting the life of the sperm to an extent which makes conception unlikely even when mating takes place close to the time of ovulation.

10

FOETAL STRESS, DEATH AND ABORTION

Pregnancy is a make-or-break event in the life of the individual. Let us take, for example, a two-year-old racehorse appearing for the first time on the racecourse, in the spring of its career. It may then be 24 months old, aged from its birth, but nearly 36 months from the time of conception. Thus, as it canters down to post carrying the hopes of its owner, breeder and trainer, one-third of its existence has been spent in utero. Pregnancy, to the racehorse at least, is a most important basis for future success. Any failure in development, damage or disease sustained during foetal life may well leave a permanent unsoundness or deformity.

Foetal health depends on the efficiency of the placenta to transfer nourishment, eliminate waste material and to act as a barrier against microbes. Usually the placental membranes perform these functions without difficulty.

FOETAL STRESS

The foetus is stressed when its environment is changed due to some abnormal happening. Let us consider some examples.

Maternal Illness

The effect of maternal illness on the foetus depends on the nature of the condition. For instance, if noxious substances or toxins accumulate in the maternal blood stream, the foetus may be unaffected because the placenta acts as a barrier. However, in certain circumstances, this barrier is not entirely successful, and poisonous material may cross into the foetal foal, causing damage to tissues and function.

The foetus of a mare suffering from fever may respond by reducing the level of its heat output and thus offset the abnormal temperature of its surroundings. If the mare's fever persists, the foetus may come to tolerate the situation and develop at this higher temperature. Much depends on the stage of pregnancy at which the fever occurs. Prolonged high temperatures can cause congenital deformities of the foetus if it is exposed to them during any of the critical stages of organ development, ie very early in pregnancy.

Infection of the Foetus

The foetus and the placenta usually contain no microbes. However, infections in the mare may cause microbes to cross the placental barrier. The most obvious example is the herpesvirus that affects the mare's respiratory system in the form of a 'snotty nose' or common cold (rhinopneumonitis). Certain types of this microbe cross the placental barrier and cause infection in the foetus, which then results in foetal stress and abortion.

External and Internal Influences

Each of the three examples quoted above relates to happenings which occur outside the foetus and its membranes. This is, of course, the same situation as we ourselves experience; microbes within our own environment challenge us from outside our bodies; poisons and injurious substances enter our bodies; and high or low temperatures affect our well-being. The difference is that our environment is composed of air and is almost limitless, whereas the environment of the foetus is bounded by the uterus and has a frontier, ie the placental membrane.

There is an important and essential difference between the direct stress suffered by the air-breathing individual and the indirect stress that affects the foetus. This may be illustrated in the following way: the placenta becomes damaged and affects the foetal environment by restricting the efficiency of the organ, thus depriving the foetus of essential nourishment or allowing the entry of toxic substances and microbes through the damaged areas. Damage to the placental membrane may be likened to our own environment becoming altered and thereby made insufficient or poisonous (polluted). A damaged placenta is a potent cause of foetal stress.

So far, we have been considering stress imposed *on* the foetus. Are there any situations where stress is initiated from *inside* the foetus? The answer to this question is probably *yes*. The development and health of the foetus depend on a normal genetic make-up, contributed at the time of fertilisation of the egg. Defective genetic material results in errors of foetal development and metabolism. We fail to recognise these cases largely because they seldom continue to full term. Those that do survive may be born with missing or deformed parts.

The Effects of Stress

The effects of stressful situations depend on the nature and degree to which the environment has been abnormally changed, the ability of the foetus to respond to the challenge, the stage of foetal development, and the duration of the challenge.

Hypoxia and Anoxia Oxygen is essential to the foetus. A prolonged (for more than three minutes) and complete absence of oxygen will do irreparable and fatal harm to the foetus. This situation we would describe as *anoxia* (literally: no oxygen), and it could arise if the mare died or if the umbilical cord became inextricably entwined in the foal's hind leg. In both of these instances the foal would die, as surely as if we ourselves were strangled. Stress situations may last for a short or long period, be severe or moderate in degree.

A more frequent happening is a reduction in the oxygen supply rather than a complete absence. This condition is described as *hypoxia* (literally: low oxygen). The placental membrane is a living organ and needs oxygen just as much as the brain, kidneys, etc. Shortage of oxygen may damage the membrane as severely as other tissues in the body. The effects on the foetus depend on the extent of damage.

Infection may be blood-borne from the maternal side, or be resident in the uterus at the time of coitus when the mare became pregnant. It may also follow damage caused by hypoxia. The infection may be mixed, as for example when a bacteria and a fungus are present. Infection may damage the placenta just as effectively as hypoxia. The degree and extent of the damage varies with the nature of the microbe and the duration of infection. However, some infections, especially those caused by virus, may cross the placental barrier; the microbes enter and infect the foetal body in the same way as they would if infecting an adult individual. Thus lesions (damage) may be found in the lungs, liver or other organs.

The response of the foetus to this challenge causes stress, in much the same way as it would to the independent individual. However, the degree of response may be muted, ie the foetus is more tolerant of the challenge. For example, periods of shortage of oxygen (hypoxia) are more readily tolerated, and therefore do less damage, than a similar degree of hypoxia would in an adult. This is an advantage to a foetus which, in any case, lives with oxygen concentrations considerably less than is necessary for existence outside the uterus. The response to infection is less advantageous because the foetus has little immunological capability to withstand invasions by microbes that cross the placental barrier.

In practical terms we recognise stress by the effects observed when the foal is born. If birth occurs before the 300th day of pregnancy we call it abortion. The foal has no chance of survival due to immaturity, and the degree of stress is often sufficiently severe to cause the death of the foetus before it is aborted. After Day 300 a foal may be sufficiently mature to have some chance of survival; and stress, in these circumstances, may cause the foal to be: (1) delivered in a premature condition, ie weak and undersized; (2) fully grown, but suffering the effects of stress in terms of weakness and difficulty in adapting to the newborn environment (see page 208); (3) unable to withstand the rigours of the birth process and be stillborn.

In summary, foetal stress is recognised by abortion, prematurity, imma-turity (dysmaturity), neonatal maladjustment, deformities, infectious disease or stillbirth. The signs are small size for breed and dates, emacia-tion (an undernourished appearance), meconium staining of the amnion and amniotic fluid, and weakness characterised by difficulty in getting up, standing and sucking. The placental membrane may be thickened or part of its surface eroded and the normal velvety surface replaced by a grossly roughened and parchment-like texture. A close examination of the placental membrane, following its expulsion from the uterus, may clearly demonstrate the degree of inefficiency which has affected the placental uterine relationship, due to a reduction in surface area brought about by damage to a substantial portion of the placenta.

ABORTION

Abortion is the expulsion or birth of the foal before the 300th day of pregnancy. People often speak of a foal having been aborted if it is born in a dead or dying condition at a later stage of pregnancy, ie after 300 days. This may be a reasonable description, but it is preferable to reserve the term abortion for birth occurring prior to the minimal period for foetal viability. Other terms sometimes used instead are 'foetal' or 'embryonic death' and 'resorption'. Foetal death is, in fact, part of the aborting process; resorption describes the death of a foetus (embryo) in the early period of pregnancy. The foetus and its membranes may become absorbed rather than expelled, and the distinction between resorption and abortion is academic. Some authorities dispute the fact that resorption actually occurs. It is not a matter of great importance to us here, because both events involve foetal death and the termination of pregnancy.

Abortions may be further described as 'early' or 'late'. The term early is, usually, applied to abortion (resorption, foetal death or foetal loss) occurring before about Day 90 of pregnancy. There is really little practical value in designating an abortion as early because we are more concerned as to *why* an abortion occurs rather than *when*. The term has been used to explain those abortions occurring after a pregnancy diagnosis has been made, often around Day 40, which is followed by a negative diagnosis made at a later date, without any evidence of abortion having been found. The evidence is missed because the foetus and its membranes are comparatively small, and the abortion frequently occurs when the mare is confined to a paddock during the summer and early autumn. In these circumstances the fact that no signs of abortion are observed may even give rise to doubts of the accuracy of the original diagnosis of pregnancy.

About fifty per cent of abortions occur spontaneously, without showing any preliminary signs. In the other fifty per cent there may be mammary (udder) development extending over days, weeks, or even months. The udder increases in size and contains an abundance of milk which, in some cases, streams from the teats in exactly the same manner as a mare at full term that 'runs milk'. This development may be associated with damage to the placenta and is particularly common prior to the abortion of twins. It also occurs in chronic infection or damage of the placental membrane.

In other respects abortion follows a similar course to foaling. But while being aborted the foal is often dead or dying and may, therefore, be abnormally presented as it enters the birth canal. This may cause some difficulty in delivery. However, the aborted foetus is usually smaller and its bones more pliable than a full-term foal, and its passage through the birth canal is usually relatively simple. The delivery may be impeded in the case of twins, if both foals are presented for delivery together. Sometimes the placental membrane may fail to separate from the uterine wall in the normal manner and has to be removed manually by the veterinarian, or the foal may be aborted with the membrane still surrounding it.

The after-effects of abortion are residual damage to the uterus causing the mare to have subsequent problems of fertility, inflammation of the uterus (metritis), and laminitis; uterine prolapse may sometimes follow.

The most important practical matter is the division between infectious and non-infectious causes (Table F, page 298). A non-infectious abortion is most unlikely to occur in epidemic proportions, whereas one caused by microbes may affect other mares. Of infectious abortions the one that is, by far, the most dangerous is viral abortion, caused by Equid herpesvirus I (EHV I: rhinopneumonitis) infection.

Non-Infectious Causes

Twins are more common in some breeds than in others and, within breeds, they may be more frequent in particular families. All species, except the horse, are capable of carrying twins to full term, and the offspring are often born healthy and subsequently grow and develop normally. Unfortunately, the position is quite different in horses. The mare has great difficulty in nourishing two foals within her uterus; and the outcome of twin pregnancies is frequently disastrous, the twins being aborted, usually between about the sixth to ninth month of pregnancy. A small number may be carried to full term but the produce are then smaller than normal and one or both of them may be born dead.

There is roughly a two per cent incidence of twin conceptions in Thoroughbreds according to returns of *The General Stud Book*. This figure has been substantially reduced due to the introduction of ultra-sound (scanning). The impact of this technique on the problem is considered fully in chapter 14. About eighty per cent of twin pregnancies abort, and twenty per cent are born at full term but with one or both twins failing to survive and, usually, much below the normal singleton size. Twins born alive at full term may fail to survive the first few days following birth and are small and weak or become unsound in later life. It is the exception which grows to an adequate size to race or breed successfully.

The reason mares cannot support the intra-uterine development of twins is that the placental membrane normally covers the whole uterine surface. So the ratio of contact between membrane and uterus is one to one. In the case of twins the two placentae compete for the available space and, at best, have to share equally the area available. In most cases, one placenta occupies more than half of the uterine wall, but it still has less space available than if it were on its own. The other twin in these circumstances has a very much reduced area available to it. The uterus may enlarge to accommodate the presence of two foetuses but cannot entirely make good the deficit. Thus both members of twin conceptions eventually suffer to a lesser or greater degree, depending on how successful each placenta has been in gaining attachment. As pregnancy proceeds, each foetus comes under increasing stress to meet its requirement for growth. Eventually one, usually the smaller of the two, dies. Although the presence of a dead foetus may not cause an abortion immediately, it adds considerably to the problems of the surviving twin by producing toxic substances. At the time of death of the first twin the mare's udder may develop, and she 'runs milk'; when eventually both twins are aborted, the

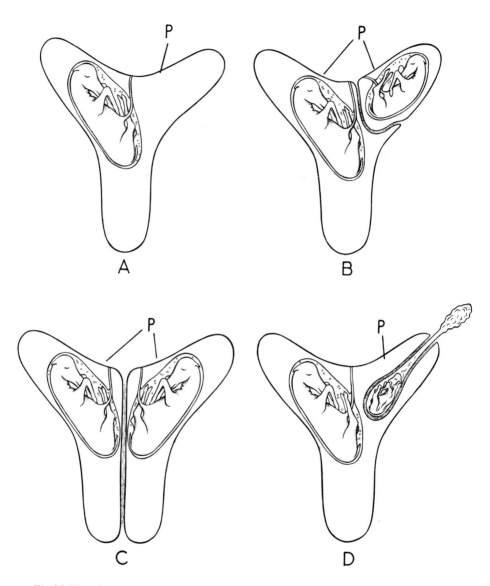

Fig 29 The placenta (P) of the singleton foal normally occupies the body and both horns of the uterus (A). If a second foetus is present each foetal foal has to share the available area of uterine wall. There are three situations which can develop: the area is shared unequally (B); the area is shared equally (C); or one member of the twins successfully walls off the other in the tip of one uterine horn (D) (From Jeffcot and Whitwell - *Journal of Comparative Pathology*)

surviving member is fresh and the first to succumb is decayed.

The females of other species are able to nourish several young within their uterus because the placental attachment is to part of the uterine wall only. Horse twins are seldom, if ever, the result of the splitting of one egg after it is fertilised. Identical horse twins do not occur, and there is always a dissimilarity in physical characteristics between each member of the pair. Horse twins are never 'freemartins' – the condition found in cattle and due to the hormones or cells of the male crossing into the female twin, causing a deformed genital tract.

Twin conceptions are a small but significant problem in Thoroughbred horse breeding. Twins are, at best, of little economic value, and at worst may leave the individual that conceives them with lasting problems of fertility following their abortion. It was once hoped that rectal examinations could be used to avoid the problem. One of the main reasons that this hope has not been fulfilled is that there are considerably more twin ovulations around the end of oestrus than are detected at rectal examinations by the veterinarian. Twin ovulations may occur as frequently as thirty per cent of the time at the height of the breeding season. A high protein content of the feed may increase their frequency but, whatever the diet or time of year, more twin ovulations occur than twins conceived.

Disturbed Foetal/Maternal Balance We have already likened the foetus to a parasite and the mare to its host. This parasite will be expelled if the delicate host/parasite balance is disturbed, which may happen in the following circumstances.

1 The placenta is damaged, thus interfering with the foetal pathways of nourishment and oxygen supply. There are many causes of damage, apart from the obvious one of infection. Perhaps the most intriguing possibility is a shortage of oxygen and/or nourishment to the placenta, which obtains its supply through diffusion from the maternal blood stream. If the placenta is deprived of its supply it becomes damaged and its role as an organ of exchange is impaired. This, in turn, imposes considerable stress on the foetus and, quite possibly, its death and abortion.

2 The umbilical cord may become entwined around the foetal hind limb, resulting in the foetus being starved of oxygen. Short periods of deprivation can be tolerated, but long periods cause death and abortion. We may liken this situation to a diver whose lifeline is partially or completely constricted. The effects on the diver will depend on the degree to which the oxygen supply is removed and for how long.

3 Interference with the proper nourishment of the placenta may result from disturbance in maternal or foetal blood flow. The heart has to propel blood through the cord into the placenta and back through the liver, thus imposing quite severe demands on its pumping capacity. Heart failure may be caused by a wide variety of adverse circumstances such as toxaemia or abnormally increased blood pressure. We know little about the effects of changes in foetal blood pressure and how these may be caused.

4 It is fascinating to speculate on how the foetus might be affected by changes in the maternal environment. Handling and transporting ponies unaccustomed to such stressful situations is known to precipitate abortion. However, we have very little evidence to substantiate many of the supposed possibilities. For example, can fright, extreme exertion, excess cooling by cold wind and rain or excess heat disturb the foetal environment, thereby producing death?

5 There is a general assumption that some abortions are caused by hormonal dysfunction. This term has no very precise meaning because of a lack of objective evidence to substantiate the diagnosis. For example, progesterone is known to be an essential ingredient of the maintenance of pregnancy (see page 148), but those levels that matter are at the site of the placenta's attachment to the uterus. But we can measure those levels only in the *mare's* blood stream; and while these reflect the concentration of the hormone in the uterine blood vessels, they are totally inadequate as a means of estimating the levels in the *foetal* blood stream, ie within the placenta. The clinician is thus ignorant of the relevant information, namely, how much progesterone is available where it is essential. We are consequently not in a position to determine if a fall in progesterone levels in the maternal blood is the result of, or the cause of, any particular abortion.

Genetic It is probable that, as in other species, a number of as yet unexplained equine abortions are due to genetic abnormalities. If the chromosomes are damaged during fertilisation, or if genes responsible for anatomical defects become incorporated into the foetal make-up, the foetus may be aborted. This represents a natural means of eliminating deformed individuals before they are born. The abortions usually take place early in pregnancy and, although they have not as yet been demonstrated in horses, we may reasonably assume that the same feature occurs in the horse as in other mammalian species.

Immunological Failure in immunological mechanisms may cause some abortions which at present are unexplained. We may marvel that a mare tolerates the presence of the 'foreign parasite' in its uterus, but the mechanisms that allow it to do so are largely unexplained. It is not difficult to envisage that the mechanism may fail in some individuals, resulting in abortion as the maternal uterus rejects the developing foetal parasite.

Nutritional The pregnant animal gives priority to the foetal needs for nourishment. Thus if the mare is short of food, the foal she is carrying may not be affected. Of course, there is a limit to the extent to which she can protect her foetus by contributing products of her own metabolism; and beyond these limits the foetus must suffer. However, even if a mare becomes completely emaciated, it is relatively uncommon for her to abort, although she may produce a small-sized full-term foal. There is evidence that the foetus may be vulnerable to specific deficiencies, especially of protein and essential amino acids, during the early stages of pregnancy. It is therefore important to ensure an adequate quantity and quality of diet from the very early stages of pregnancy, with the emphasis on an increasing availability during the last half of gestation, when the foetus is increasing rapidly in size.

Substances which should be avoided as potential abortifacients, ie drugs causing abortion, include luteinising hormone (LH) during the first six weeks of pregnancy, cortisone, prostaglandins, oxytocin and diethylstilboestrol. Some plants contain alkaloids, glyocides, saponins and photodynamic substances which may be toxic to horses and directly or indirectly cause abortion. These include ragwort, horse tails, monkshood, bracken, greater celandine, poppy, flax, cowbane and bryony. Ergot is a substance which is found in rye and other grasses and may stimulate the uterus to contract and thus cause abortion.

Infectious Causes

Virus The most important virus causing abortion is Equid herpesvirus I (EHV I). EHV I was once divided into subtype 1, causing abortion and respiratory symptoms, and subtype 2 causing respiratory signs only and rarely abortion. It is now customary to describe subtype 1 as EHV 1 and subtype 2 as EHV 4. The infection known as rhinopneumonitis or viral abortion was given its name by workers in Kentucky, where the disease was first recognised. However, the virus has been classified by modern techniques as belonging to the group *Herpes,* among whose members are

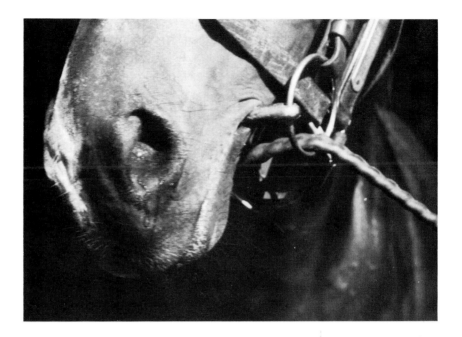

Nasal discharge may be serous (watery, above) *turning to muco-purulent (thick,* below)

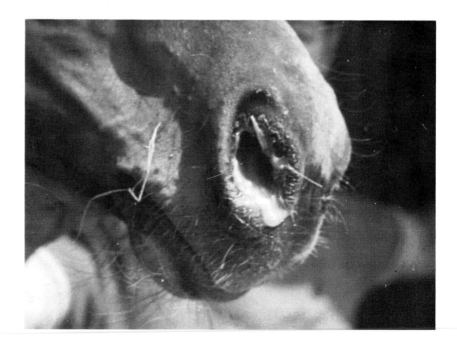

those causing diseases in man including herpes simplex sores and shingles. The equine virus is, of course, quite distinct, and the disease it causes is essentially one of the respiratory tract. Abortion is an incidental symptom only.

Symptoms of the respiratory infection include a nasal discharge especially affecting young horses under the age of four. The discharge may be watery or become thick and purulent. Affected individuals may suffer slight fever, but they generally show little or no malaise, apart from the occasional cough, a runny nose and eyes. The symptoms are the result of inflammation of the mucous membranes lining the airways in the head and lung: hence the name rhino (nose) -pneumon (lung) -itis (inflammation of).

The infection is highly contagious in foals and yearlings and, like the common cold in humans, it runs its course lasting weeks or even months, until eventually the individual establishes some immunity and resistance, throwing off the symptoms until the next attack occurs. In racehorses the symptoms may be rather more serious than in unbroken horses because athletic performance may be affected. The virus may also cause a chronic pharyngitis.

Certain strains of Equid herpesvirus are more likely to be associated with abortion than are other strains; the terms 'abortion' or 'respiratory' strains are used by laboratory workers. The situation is complicated by the variation in immunity which influences individual susceptibility to abortion, when a pregnant mare is challenged by either a respiratory or an aborting strain. Vaccines have been developed and are used extensively in many countries, including the UK. In recent years we have witnessed two significant developments in the pattern of rhinopneumonitis cases, namely the increase in the number of full-term pregnancies that are affected and the outbreak of a paralytic form of the disease affecting mares, stallions and foals such as that which first occurred in the Newmarket area in 1979.

The classic form of rhinopneumonitis is a sudden unheralded abortion, most commonly occurring in the seventh to ninth month of pregnancy. Another feature is the respiratory attack (runny nose) preceding the abortion, sometimes by several weeks. Abortion often occurs singly, but in many instances more than one mare is affected; sometimes abortion storms take place in which the majority of mares on an affected studfarm abort within a period of some four to eight weeks. Originally the condition was recognised as being highly infectious long before the causal microbe was discovered. It was considered most probable that the virus spread from the respiratory tract of one infected mare to another, and that the virus remained in the mare's tissues until such time as it crossed the placenta to infect the foetus and cause abortion. The pathology of the foetus is fairly characteristic, namely an enlarged liver with pinhead-size necrotic nodules (foci) spread throughout its substance. Destruction of the liver cells, causing jaundice and damage to organs, including the lungs, heart and blood vessels, results in effusions of fluid in the abdominal and thoracic cavities. Another special pathological feature is the presence of so-called 'inclusion bodies' in the nuclei of liver cells, around the edge of the 'microabscesses'.

The disease, in its present form, is much the same as it was reported in the 1930s except for two notable changing patterns. Firstly, nowadays we do not recognise, in most cases, preceding evidence of colds or snotty noses in affected mares. And secondly, the abortion may come at the end of a full-term pregnancy involving the delivery of a live foal suffering from signs of infective disease (see page 297), ie sleepiness and a loss of the ability to get up and seek the mare to suck.

During the last decade research workers have found EHV 1 infection to exist in a *latent* form. Latency entails the presence of virus in body cells without activity involving virus replication or symptoms. Stress or other factors may cause the latent virus to become active, months or years after the initial infection. This is called recrudescence.

Bacterial It is interesting and probably significant that the bacteria associated with abortion are the same as those encountered in the barren uterus, which, as we discussed in chapter 3, cause infertility. These are the same organisms, with few exceptions, that cause illness and fatalities in newborn foals. The late Murray Bain, a distinguished vet from the Scone Valley, Australia, drew attention to this sequence of microbial infection and showed that infections at the time of conception were often carried over into the pregnant and newborn period.

Bacteria may infect the placenta and/or the foetal body. Infection of the placental membrane may, in itself, cause abortion, and it is not essential for the organism to penetrate to the foetal body and its organs. However, pathologists face the diagnostic problem of how to interpret the presence of microbes in the foetus following abortion. *Streptococci,* for example, are common inhabitants of damaged tissue; therefore their presence could be the result of, rather than the *cause* of, foetal death. The foetus and its membranes are normally sterile, but bacteria may gain entrance as a result of blood-borne infection in the mare, similar to the route taken by the virus; or microbes may be present in the uterus at the time of conception, meeting the embryo on its arrival in the organ after its descent down the Fallopian tube. In some instances, microbes may enter through the cervix during pregnancy. Whatever route is taken by the microbes they may not establish a true infection in the foetus, and thus may be tolerated until the end of pregnancy. However, if a true infection is established, ie one causing damage, the illness from which the foetus suffers is similar to those caused by viral or bacterial infection in the body of the newborn or older foal. For example, the organs are damaged by direct action or, indirectly, by toxin produced by the microbe.

Fungus (Mycotic) Abortion Fungus, in contrast to virus and bacteria, is slow-growing and persistent. Fungal spores may, as we have already discussed, enter the uterus after foaling or in the covering yard at the time of coitus. Their growth may be enhanced, prior to conception, by the use of antibiotic treatment, destroying bacterial flora but allowing fungus to flourish. The fungus does not always reach the foetal body but slowly erodes the surface of the placenta, and eventually deprives the foetus of essential nourishment, causing interference in its growth and, in many cases, its abortion. The area most frequently affected is the cervical pole of the placenta.

CONTROLLING THE SPREAD OF INFECTIOUS ABORTION

Viral infection can cause abortions on an epidemic scale. When an abortion does occur, therefore, we should treat it as a potential case of virus infection, until a diagnosis can be established to the contrary. The problems of diagnosis and the length of time necessary to ensure accurate results are considered later. We should also remember that the rhinopneumonitis virus can cause disease of the newborn foal; and the precautions required to control the spread of infection following abortion apply equally to illness in the newborn period.

The principles of control are as follows.

1 The dead foetus (stillborn or dead newborn foal) should be placed in a leakproof container, such as a strong plastic bag. The afterbirth (foetal membranes) should be collected in a similar manner, and the two containers removed to a suitable veterinary laboratory for examination.

2 The aborting mare should be isolated, as far as possible, and kept out of

contact with any other pregnant mare. The best method of achieving this isolation is to leave the mare in the loose box in which she has aborted and maintain at least one empty stall on either side. Attendants should disinfect their boots and wear special protective clothing on entering the affected mare's loose box. If practical, one attendant should be detailed to look after her requirements, including mucking out and feeding. This attendant may perform duties with barren mares but should wear protective clothing that can be disinfected and changed outside the loose box so as to avoid carrying the virus from one part of the stud to another. Adequate facilities for washing hands and disinfecting boots should be provided close to the loose box where the affected individual is stabled.

3 If a mare aborts in a paddock she should be brought into the stabling in which she is regularly housed or which is convenient for the purpose of isolating her. The most important aim is to avoid bringing the mare into indirect or direct contact with any pregnant mares, especially those with which she has not previously been in contact. In this context mares in *direct* contact are those which have been stabled in the same stable range or grazed in the same paddocks during the previous two months. *Indirect* contacts are those mares which have not had direct contact, but which have been on the same premises, handled by the same personnel or transported in the same horsebox as the affected mare during the previous two months.

4 No pregnant mares should be moved onto or off the studfarm for at least one month following the abortion, and preferably not until all mares have foaled. Barren and maiden mares can be brought onto the studfarm for mating if no further abortions occur within a period of four weeks. In principle, however, the less traffic into and out of the stud the better because of the unknown incubation period between infection and abortion (fig 30).

5 A mare that has aborted her foal is considered safe for mating one month after the abortion.

6 It is good practice to put all placentae in a plastic bag immediately after foaling and to use several foaling boxes (see page 191).

7 All the above precautions should be introduced at the time of an abortion or the death of a newborn foal, but they may be lifted when the postmortem has confirmed that death was not due to virus.

DIAGNOSIS

Diagnosing the cause of abortion is a matter for the pathologist. The foetus and its membranes may be examined with the naked eye, followed by laboratory examinations and tests. Gross damage to the placenta may be recognised by an abnormal thickening and/or the colour and nature of the surface. This is not the place for a detailed pathological description, but the reader who wishes to become informed in these matters should

LAYERS OF ISOLATION WITHIN A STUDFARM OR STABLE

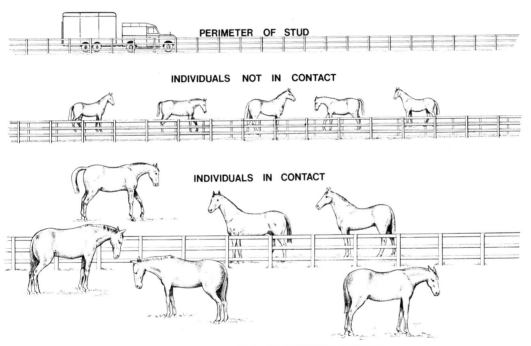

PERIMETER OF STUD

INDIVIDUALS NOT IN CONTACT

INDIVIDUALS IN CONTACT

INDIVIDUALS WITH SIGNS OF INFECTIVE DISEASE

Fig 30 The 'layers' of isolation to reduce the risks of epidemic infection spreading. Individuals with signs of infective disease should be isolated together. A second layer of isolation involves those individuals directly or indirectly in contact. Individuals not in contact but on the same premises should be isolated as far as possible. No animal should move onto or off the studfarm until diagnosis has been satisfactorily established and the question of the likelihood of further cases resolved

closely examine the placentae of normal gestation and birth and achieve a good knowledge of the normal, so that a diseased membrane may be more readily recognised.

Fungus causes a gross thickening of the placental membrane, the uterine surface of which is covered by a thick sticky brown discharge. In streptococcal infections, and where the membrane has separated from its attachment to the uterine wall, the surface may be smooth and brown or white and parchment-like compared with the normal red/purple velvety surface. In some cases the membrane adheres to the horns of the uterus, preventing the afterbirth being detached following the abortion.

The amnion, too, may be grossly abnormal, thickened and discoloured. The amniotic fluid and amnion may be stained brown with meconium. If the foetus suffers from a shortage of oxygen, meconium is expelled owing to intestinal spasms; so it is frequently to be found in cases of abortion.

The foetal body may be fresh, and many foetuses are aborted with their heart still beating, although quite incapable of surviving or establishing a breathing rhythm because of their extreme immaturity. In other cases, the foetus may have been dead for hours or days prior to being expelled, and the tissues are degenerating. The pathologist recognises other differences, such as the state of development (presence or absence of hair, length and weight of body), relative to the gestational age. A comparison with normal values provides evidence as to whether the placenta was functioning normally up to the time of abortion. If the placenta was damaged some time before the foetus was aborted we would expect to find that foetal growth had been retarded. On the other hand, if placental function was normal up to the time of death the foetus would be fully sized and developed for its gestational age.

A full account of the methods of diagnosing the cause of abortion is not appropriate in a book for lay readership. However, it is important that those responsible for brood mares should understand the process by which a diagnosis is made and, particularly, the problems facing the pathologist in establishing the diagnosis of viral abortion. We have already seen how important it is to regard every abortion as potentially one caused by virus, and to implement appropriate precautions until the case is proved otherwise. The precautions outlined are quite demanding because they entail a clamp-down on movement within, into and out of, the studfarm. At the height of a busy breeding season this obviously imposes a somewhat harsh discipline. Management requires an accurate and quick diagnosis to be made; but, unfortunately, speed of reporting and accuracy are sometimes conflicting commodities.

The first step in diagnosis is to open the body and to examine the internal organs. There are certain features which make us suspect that virus may be present. These include a jaundiced appearance of the carcass, a yellowish or bloodstained fluid in the abdominal and thoracic cavities, an enlarged liver containing pinhead-sized brown spots on its surface and in its substance, and lungs appearing mottled on the cut surface. Any one of these signs leads to suspicion, but not the certainty, of viral infection; their absence does not necessarily exclude such a diagnosis.

Specimens of liver may be examined under the microscope for typical appearance of damage caused by EHV I. This can be recognised by the multiple small areas of cell destruction with the presence, in the nucleus of the cells at the periphery of the area of destruction, of inclusion bodies. These may also be found in the cells of other organs, such as the lung and thymus. These tissues are examined within hours of receipt at the laboratory, by a special process known as 'frozen section'. The results are not 100 per cent reliable, and confirmation of findings – negative or positive – must await an examination of the tissues put through standard but more lengthy methods of staining and preparation, taking up to three days.

Viruses are grown by special methods involving tissue culture (the inoculation of virus into living cells). This process may take up to two weeks before it can be proved that a particular virus was present in the foetus. This culture test is the deciding factor of the diagnosis, so conclusive evidence of a viral abortion may take two or even three weeks.

In summary, the diagnosis of virus abortion depends on the gross appearance of the foetus and its membranes, the examination of a frozen section of the liver, followed by a search for the presence of inclusion bodies in the liver and other organs and, finally, on cultivating the virus in tissue culture. A negative diagnosis is based on the absence of any pathological signs suggesting the presence of virus together with positive evidence of other causes. However, in assessing the situation, we must be careful because, for example, although twins are an obvious cause of abortion it is also possible for these to be affected by virus.

The final judgement is, of course, in the hands of the veterinarian and the pathologist, whose responsibility it is to advise the client in any particular circumstances. The client should have sympathy for the problems which face the pathologist in making the diagnosis, be patient in waiting for the answer and be prepared to take precautionary measures until such time as the investigation is complete. It is better to be safe than sorry, especially when to be sorry may mean the unnecessary spread of a very serious disease.

The diagnosis of other forms of abortion is made along similar lines – a jigsaw puzzle where the pieces (results of gross and microscopic appearance, bacteriological and virological cultures) are painstakingly collected and put together to form a picture. The biggest obstacle to success is the state of the material when it arrives at the laboratory. The reader should appreciate that a foetus that dies in the uterus may be retained for some period at the body temperature of the mare. This is equivalent to keeping meat in a room at 100°F (38°C), and no one would expect it to remain in prime condition under these circumstances for very long. The same applies to the condition of a foetus. Putrefaction starts within hours of a foetus dying; the chances of obtaining suitable material for laboratory examination decline correspondingly. The diagnostician is thus attempting a task which is quite unique in medicine, namely to deduce the signs and process of disease without either examining the 'patient' at the time of its illness or being presented with satisfactory material for post-mortem examination.

It may be helpful to examine the mare for evidence of infection and perform tests on her uterus, such as obtaining biopsy specimens, to help in the diagnosis of the cause of an abortion.

MANAGERIAL INFLUENCES ON ABORTION

Every horse owner wants to know how to prevent or avoid the disaster of an aborting mare. However, veterinary science cannot offer precise answers to many of the questions which spring to the horseman's mind concerning the avoidance of abortion. A great deal more investigatory work is required before we can provide the answers to such questions as: Is there an optimal time in pregnancy for transporting pregnant mares? Is it dangerous to wean (or leave unweaned) the foal at foot beyond a defined stage of pregnancy? Can the management of a mare cause abortion?

It may be helpful to list *possible* harmful effects of management, some of which have already been mentioned. Sudden and extreme changes in management are the most probable risks, whether this involves changes in handling (eg from being at pasture to being confined in loose boxes) or changes in diet. It has recently been shown that withholding food for 12 to 24 hours causes a rise in prostaglandin levels in the mare, and this hormone may precipitate abortion. More specifically, one might envisage problems being caused at weaning time, in the autumn, when mares are placed on low-quality diets, having experienced high-quality feed and being housed at night, while they had their foals at foot. Other shocks to the system include being subjected to heavy and prolonged rain, together with cold winds causing severe chilling. Upsets in routine, fright caused by loud sounds, and other physical disturbances might be expected to have adverse effects. In the past, low-flying aircraft have been blamed for causing abortions, as have packs of hounds running close to paddocks in which mares are grazing.

However, it must be stressed that there is no substantial evidence to designate any of these happenings as a definite cause of abortion. It seems more probable that some individuals are susceptible and therefore more likely to abort than others. The susceptible individual may lose her foal whereas mares that have no such susceptibility will remain unaffected, except in the most extreme circumstances. Nevertheless, the advice to management must be to shield all mares from abrupt changes in management, from adverse climatic conditions and nutritional imbalances or dramatic changes in diet. Further, pregnant mares should always have easy access to clean drinking water.

Preventive therapy, such as administering progesterone or antibiotics during pregnancy, comes within the province of the veterinarian, and readers are advised to consult their own vets in these matters.

11

DIFFICULT FOALING

Foaling is a natural event, but it has all the appearance of drama – delivery is rapid, the mare's efforts are extremely forceful, and it is a unique biological occasion, the separation of two individuals. The inexperienced observer may, in consequence, be fearful of the outcome, attempt to hurry or interfere with the process unnecessarily and imagine that the mare is incapable of completing the safe delivery of her foal. In fact the majority of mares have no difficulty in foaling, although some problems are more common in certain breeds, such as Thoroughbreds, than in others, such as ponies.

Difficult birth (dystokia) is often regarded solely in terms of problems of delivery in which the foetal foal becomes lodged against the mare's pelvic girdle. However, in this chapter, we must consider the broader implications of dystokia, including abnormal conditions of first- and third-stage labour and the effects of birth on the foal as well as on the mare.

Abnormal First-stage

Abnormal first-stage labour is usually the result of the thickening of that part of the placental membrane adjacent to the cervix. This, the cervical pole, is normally thin and easily ruptured as the cervix dilates and the foal's limbs press against it, just before the onset of second-stage. The membrane may be abnormally thick because of infection, oedema (water-logging) or hypoxic (shortage of oxygen) damage, and it is too strong to be broken by the normal forces of birth. Thus it is carried through the birth canal and appears unbroken at the vulval lips, after what may be a prolonged first-stage. The inability to rupture the membrane in the normal manner may cause the mare some discomfort, and instead of waiting to break water in the standing position she may lie down and strain. This action pushes the membrane through the vagina and vulval lips and, if no assistance is provided, it tears further back so that the cervical pole separates completely. Studmen should break the placental membrane if it appears at the vulval lips. It may be recognised by the star-shaped scar present in the centre of the bulging membrane. This scar may be used to

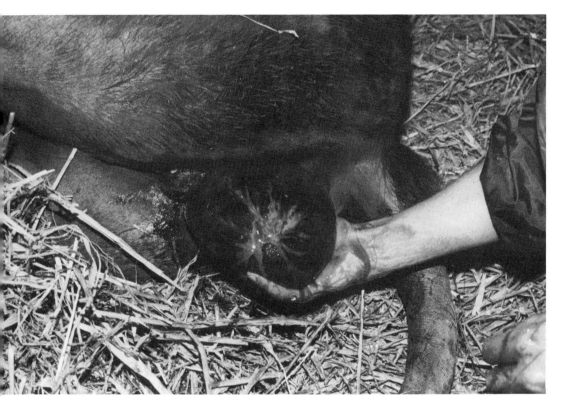

Placental membrane (held in hand) *appearing abnormally at the end of first-stage labour; note the scar which can be used to distinguish the membrane from a prolapse of the bladder or uterus. The membrane should be broken with the hand or a pair of scissors*

identify the membrane and distinguish it from the unlikely event of a prolapse of the bladder or wall of the vagina. It may be quite difficult to break with the fingers and scissors may be required. Once the membrane has been broken, second-stage may proceed normally.

Abnormal Second-stage

Malalignment of the foetus is a common cause of difficult delivery. The foal may be presented wrongly, lying awkwardly relative to the bony surroundings of the birth canal and/or with an abnormal posture of the head, neck and/or limbs. Two foetal malalignments are illustrated in figs 31 and 31a.

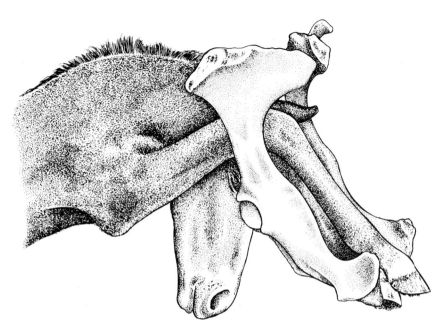

Fig 31 & 31a Abnormal postures: foal entering birth canal with head obstructing progress

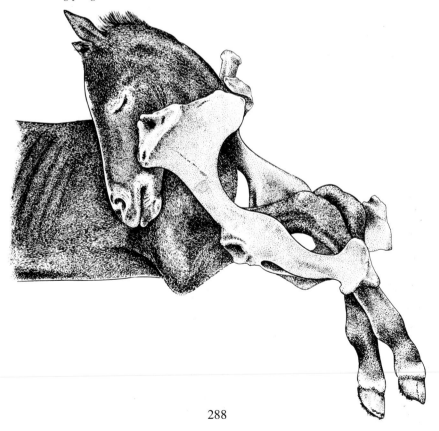

Foals that are coming back to front, ie hind feet first (posterior presentation), may be diagnosed during early second-stage by identifying the hocks and tail in the birth canal. This type of presentation is very abnormal and unusual. It does not necessarily impede delivery, because the foal can usually slide backwards quite easily through the birth canal. However, passage in this manner drags the umbilical cord over the mare's pelvis and it becomes squeezed between the chest of the foal and the bony surface below, thus cutting off the foal's oxygen supply while its head is as yet undelivered. Foals delivered backwards should, therefore, be pulled from the mare as rapidly as possible once the hocks have emerged beyond the level of the vulval lips. Even if this action is taken, the foal is likely to be born dead or badly affected by asphyxia.

Malposture of the forelimbs and/or head and neck may be suspected if any one of these members is not present at the examination made early in second-stage (see page 172). We may expect, for example, to find the two forelegs but no muzzle in cases where the neck is flexed to one side. These postural abnormalities prevent the foetus from entering the birth canal, and we may suspect their presence where mares do not proceed with delivery following the breaking of their waters. The mare's behaviour may also indicate malposture if she either fails to strain or gets up and down repeatedly, as if attempting to shift her foal into the correct alignment.

The difficult delivery caused by malposture of the head and forelimbs does not put the foal's life immediately at risk, because the cord is under no pressure and the placental membrane is still attached to the uterine wall. Thus the foal's lifeline is secure, until such time as the abnormalities can be corrected and the foal delivered in the normal way.

On the other hand, abnormal posture of the hind limbs may cause the hind feet to lodge on the pelvic brim. By this time the foreparts and chest of the foetal foal are engaged in the birth canal, and the cord is in danger of becoming squeezed. This condition is diagnosed by the fact that the foal, having advanced to the stage shown in figs 32 and 32a, makes no further progress, even if a strong pull is exerted on the foreparts, in an effort to complete delivery. In fact strong traction in these cases may appear to be moving the mare rather than her foal. In this situation the foal's lifeline is seriously at risk, and we have little time to correct the difficulty. The more we pull the more firmly the hind feet become lodged on the pelvic brim, so we must advance our hand along the floor of the birth canal and attempt to push the foetal feet (or foot) off from the pelvic brim. Another manoeuvre is to twist the foal one way and then the other,

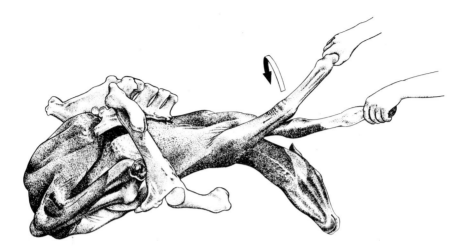

Fig 32 & 32a The hind feet have become lodged on the brim of the pelvis; the foal, at this stage, is nearly delivered. The only way of dislodging the foot is to pass a hand along the floor of the vagina and/or to twist the foal as illustrated. This method causes considerable pressure on the foal's chest and it is advisable to supplement the rotational movement by having a second person grasp the skin on either side of the chest and shoulder region

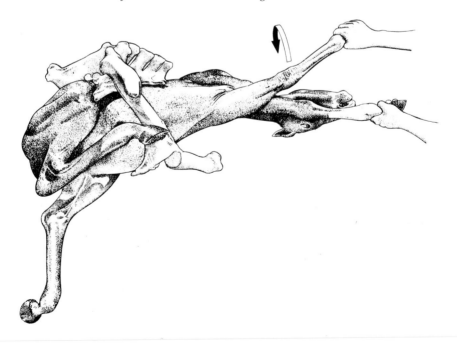

in an attempt to dislodge the feet by indirect means. The advantage of this method is that it will deal with the alternative diagnosis when the foetal hips and not the hind legs are engaged against the maternal pelvis. Great care must be taken in rotating the foal. The chest is particularly liable to damage if we twist the body while the hind parts are lodged, and we may readily fracture the foal's ribs by our action. One of the best means of turning the hind parts of the foal is to strip the amnion away from the foetal body and grasp the skin on either side of the flanks. A second person should then take the chest and support it as the foal is rotated first one way and then the other, pulling gently at the same time towards the direction of delivery, ie downwards towards the mare's hocks.

The position of the foal as it enters the birth canal should be as described on page 181: the backbone of the foal passes slightly to one side or the other of the mare's spine. In other words, the foal's withers and chest should pass through the pelvic hoop in a plane slightly diagonal to the vertical plane of the mare, ie slightly to one side of the line formed by the vulval lips. The foal's shoulders may become lodged against the mare's pelvis if its position is too upright or too greatly inclined, ie lying too much on its side. The reader will remember that the foal starts first-stage in the position of lying on its back and rotates during first- and early second-stage. If this rotation has not been completed, the mare is unable to push the foal through the birth canal.

Major manipulations of the appendages or of the foetal body should be made with the mare in the standing position, because the foal is then drawn back into the uterus under gravitational force. This allows more room for manoeuvre in, for example, readjusting the disposition of the foetal limbs. The mare's straining efforts have an opposite effect, reducing the room for manoeuvre, straining and pushing the foetus into the birth canal, where there is insufficient room to correct abnormalities such as a flexed forelimb. It may be necessary in certain circumstances to administer some form of spinal anaesthesia or drugs which help to reduce or abolish straining. This is, of course, a matter of veterinary judgement, and if there is a need for this type of control it is imperative to call for skilled assistance at an early stage. The longer a mare is in labour without foetal malalignments being corrected, the greater the risk that even the most skilled operator will be unable to succeed in the task of realignment. The mare's unavailing efforts of straining soon produce swelling and dryness of the birth canal, making manipulation more difficult, or even perhaps impossible. There are further risks of the placental membrane separating from the uterine wall, causing foetal death and/or shock, and possibly uterine rupture in the mare.

Once the decision has been made that delivery is impossible without skilled assistance, the mare should be kept standing or walked around the loose box while awaiting the arrival of the vet.

Rectovaginal Fistula We have already discussed the duties of studmen at birth (page 195). One hazard of birth is if the foal's forelegs are driven through the roof of the vagina into the rectum. This may happen if the mare lies down at the time when the front feet are lying in the vagina and the foal's body is angled so as to direct the forelegs in an upward direction. A mare that has been stitched may be particularly susceptible to this occurrence because the stitched part of the vulva may form a flap, preventing the forefeet from passing readily through the vulva. If it appears that this accident is about to occur, those present should guide the forelegs away from the roof of the vagina and through the vulva.

Immediate action should be taken if the foal's forelimbs or head are seen emerging from the anus rather than from between the vulval lips. The mare should be made to rise to a standing position and the appendages pushed back into the birth canal through the tear between the rectum and vagina. If this can be achieved the foetus can then be guided through the normal channel of the birth canal, leaving the fistula between the vagina and the rectum to be repaired surgically after birth.

Foetal Deformities Before leaving the subject of difficult delivery, mention should be made of one common cause, namely deformities of the foetal appendages or body, especially those involving contraction of the forelimbs. This may make it difficult or impossible for the foal's forelimbs to be sufficiently extended to enable them to pass over the brim of the maternal pelvis into the birth canal. The foal will probably present with its forelimbs flexed at the knees and only the foetal head entering the birth canal. Even skilled veterinarians may be unable to bring the knees into extension, and it is then necessary to resort to methods such as embryotomy (cutting off parts of the body, eg the forelimbs) while the foal is still in the uterus, or delivery by caesarean section. In some instances it may be possible to deliver deformed and badly malpostured foals through the vagina under general anaesthesia, without recourse to surgery. The decision in these matters is, of course, made on veterinary grounds, backed by a knowledge of the circumstances in which foaling is taking place.

Uterine Haemorrhage

Haemorrhage (bleeding) associated with foaling may occur in two forms. The most common and dangerous type is haemorrhage from a break in one of the major arteries supplying the uterus or vagina. Because this is an internal haemorrhage the blood does not usually escape to the outside. The symptoms are those of pain – the mare rolling violently and getting up and down repeatedly as if with colic – followed by signs of massive blood loss including a staggering or swaying gait and, depending on the amount of blood lost, fainting, shock and death. The haemorrhage occurs during or within two days of foaling, and most commonly in older mares. Other signs include paleness of the visible mucous membranes (mouth, eyes and vagina), which turn yellow over several days as jaundice develops. A swelling may appear on one side of the vulval lips as the haemorrhage tracks backwards along the side of the uterus and vagina. In some rare cases the haemorrhage may actually bulge into the wall of the vagina, breaking through the wall and flowing in large quantities through the vulval lips.

The blood vessels are carried to the uterus in the supporting membrane, the broad ligament. If one of the blood vessels breaks, blood escapes into the membrane, stretching its wall – this accounts for the symptoms of pain. The haemorrhage may be contained within the membrane, and in these cases the condition becomes less painful as recovery proceeds. The anaemia and jaundice are the result of blood loss and the breakdown of the massive clot that forms in the broad ligament. If the blood breaks through the membrane, into the abdominal cavity, there is nothing to stem its flow, and the mare bleeds rapidly to death. This usually takes place immediately after delivery of the foal.

A haemorrhage contained within the broad ligament may be felt by the vet *per rectum* as a large mass several days after foaling. This finding confirms the original diagnosis based on signs of pain, jaundice and anaemia.

The second form of haemorrhage associated with foaling is bleeding from the uterine wall, that is, into the uterine cavity; this is not usually serious, because it is not nearly as extensive as the internal bleeding described above. In fact much of the blood passed by mares after foaling has come from the placenta, ie foal's blood mixed with a small amount of mare's blood that has oozed from the wall of the uterus. The important point to understand is that this uterine bleeding in the mare does not compare with uterine haemorrhage in other species, because the attachment of the placenta to the mare's uterus is such as to cause minimal damage when it peels away after foaling. In women, the placenta is joined to the wall of the uterus in such a way that it brings away maternal tissues when it comes detached; in the mare, little, if any, maternal tissue comes away with the placenta.

Recovery of the Uterus after Foaling

Vets speak of uterine involution. This term describes the return of the uterus to the size and condition of a barren mare. It is remarkable how quickly this occurs: to a large extent by the foaling heat, nine to fifteen days after foaling. Thus the uterine walls shrink from a condition of maximal distension, immediately before the foal is born, to a relatively small size within a fortnight; and the lining undergoes similar dramatic changes from the pregnant to the non-pregnant state. Of course, the recovery may take longer in some individuals than in others and, as we saw in chapter 9, complete recovery may take several weeks, months or even, in exceptional cases, years. Failure of the uterus to recover (involute) may complicate the immediate post-foaling period. The signs include a copious discharge of bloodstained fluid from the vagina, symptoms of mild pain and unease.

Retained Placenta

The afterbirth normally comes away from the mare within an hour of the foal's delivery. The membrane peels away from the uterine wall, the process starting at the uterine horns and continuing down the body to the cervical region, by which time separation is complete. The process is based partly on a biochemical system involving enzymes and hormones, the reduction of blood pressure in the microcotyledons (see page 181) causing them to release their 'grip', the pull of gravity (weight) of the extruded portion of the membranes hanging down behind the mare, and partly on an active expulsive process brought about by the contracting uterine wall. These mechanisms may fail and the placenta is left attached to the uterine wall for an abnormal period. Nowadays most vets define this abnormal period as being more than ten hours and, at this time, they advise manual removal. It was once customary to make every effort to remove all of the attached placental membrane; it is now usual practice to remove most, but not necessarily all, of the membrane. The tip of the placenta, which is the most difficult to remove, may be left for several days, when it comes away quite easily. The firm attachment of one horn of the placenta to its corresponding area in the uterus is often the cause of retention. Veterinary therapy of retained placenta may differ with individual experience, but most treatments are based on administering the hormone oxytocin which causes the uterus to contract.

Other Complications of Foaling

Uterine prolapse is a condition which may be encountered in young mares foaling for the first time and, less frequently, in older mares. The uterus turns inside out, so that its surface lining protrudes through the birth canal and the organ comes to hang behind the mare. It is, again, a veterinary responsibility to deal with this condition. It is sufficient here to suggest to

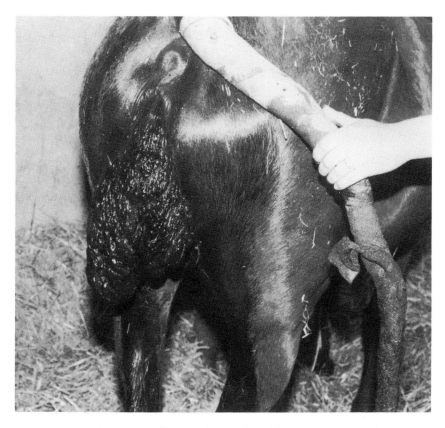

Uterus extruded through vulva (uterine prolapse) in pony mare who has recently foaled. This was replaced successfully and the mare got in foal during the same breeding season

readers that in the event of this unfortunate happening, they should (1) immediately seek veterinary assistance, (2) while waiting for this to arrive, keep the mare in a standing position, and (3) attempt to support the protruding organ with a clean, damp cloth or sheet to protect it from further damage caused by drying and by contact with straw or hard objects. Normal saline, ie one teaspoon of salt to a pint (0.6 litre) of water, is the ideal solution in which to bathe the surface of the organ.

Rupture of the uterine wall is a rare complication of foaling. It should be suspected if the mare fails to strain during delivery or shows excessive afterbirth pains. Its diagnosis is obviously a professional matter and it has to be distinguished from other conditions causing similar signs, for example rupture of the caecum or colon. These rare complications are caused by the forces of birth and/or by the foetal appendages striking against the adjacent organs.

Fracture of the pelvis or femur (thigh bone) may sometimes result from the mare 'doing the splits'. The most likely moment for this to happen is when a mare gets to her feet with her foal half-delivered, ie with the chest emerging from the vulval lips. The weight of the foal, in this situation, may cause considerable difficulties in balancing as the mare rises to her feet, causing her to fall with her legs straddled apart. Once such a fracture has occurred the mare is unable to rise to her feet and will probably have to be destroyed to prevent further suffering.

12

DISEASES OF THE NEWBORN FOAL

The newborn foal is a peculiar being in the eyes of veterinary clinicians. It is, in effect, a patient with no known history. In all other medical situations we can observe the patient at the onset of its illness, perform a physical examination when first signs appear, and learn of the circumstances leading up to the illness at first hand from those in attendance. The newborn patient is, more often than not, suffering from illness the origins of which occurred prior to or during birth. In the newly born, we can gain information from examining the placenta, amnion and amniotic fluid; and we usually know something of the manner of foaling, eg if it has been prolonged or difficult. Apart from this, vets have to make the diagnosis on present rather than on past knowledge.

Vets do, however, have one important advantage – namely that of knowing that foals suffer from specific conditions related to this unique period of life; they also know that the signs of this disease will, with few exceptions, first appear within the first four days of life.

CLASSIFICATION

It is convenient to classify foal diseases into four groups and to distinguish between infectious and non-infectious conditions. *Group1* comprises infections caused by a large variety of microbes (Table F); *groups 2, 3 and 4* include non-infectious conditions, each based on broadly differing natural types of disease, with sufficient dissimilarity to put them into separate categories.

In *group 2* conditions (Table G) we place such well-known diseases as prematurity (gestational age less than 320 days), dysmaturity, convulsive foals (the neonatal maladjustment syndrome, NMS), and meconium colic.

Group 3 conditions are anatomical abnormalities brought about by adverse influences on the foetal body during pregnancy. The more common defects are 'parrot jaw', contracted limb tendons, hernias, absence of one or both eyeballs, and cleft palate. We have little knowledge of the cause of these defects in horses, but from a study of other species, including humans, we may deduce that the main reasons are:

TABLE F Infectious conditions of the newborn foal placed in group 1

Condition	Synonyms	Characteristics	Cause
Generalised infections	Diarrhoea Scour Pleurisy Pneumonia Peritonitis	Fever to 102°F (39.5°C) and persisting Diarrhoea Increased respiratory rate, with rales Dehydration, retraction of eyeballs	*Escherichia coli* *Streptococcus* spp
Hepatitis	Virus abortion Rhinopneumonitis	Convulsions Lethargy Mild jaundice	*Equid herpesvirus I* *Cytomegalovirus*
Nephritis	Viscosum Shigellosis Sleepy foal disease Sleeper	Initial fever to 102°F (39.5 °C) becoming subnormal Sleepiness Diminished strength of suck Diarrhoea, mild colic Convulsions Uraemia: protein, cells in urine	*Actinobacillus equuli (BVE)*
'Meningitis'	Convulsions	Fever to 102°F (39.5°C) and persisting Convulsions	*Streptococcus* spp *Escherichia coli* *Actinobacillus equuli*
Encephalitis		Gross disturbances of behaviour	
Infective arthritis	Joint ill Navel ill	Fever to 102°F (39.5°C) and persisting Lameness	*Streptococcus* spp *Salmonella typhimurium*
Tenosynovitis		Painful swelling around joints	*Staphylococcus* spp *Escherichia coli*

From *The Practice of Equine Stud Medicine* (Rossdale & Ricketts, 1974)

TABLE G Non-infectious conditions included in group 2 characterised by gross disturbances in behaviour

Condition	Synonyms	Characteristics	Cause
Neonatal maladjustment syndrome (NMS)	Barkers Wanderers Dummies Convulsives	Full-term gestation First signs within 24 hours; often normal until onset Complete loss of suck to nurse Apparent blindness Convulsions, spasms Coma Fever	Asphyxia Birth trauma Foetal stress
	Respiratory distress	Fast and distressed breathing	
Prematurity		300–320 days' gestation Signs apparent at birth Weakness, delay in first standing Low birth weight Reduced strength of sucking, ability to suck and ability to maintain body temperature Tendency to colic after feeding 'Bedsores' Discoloured tongue, silky skin	Foetal stress
Immaturity	Dysmaturity	More than 320 days' gestation Appearance and signs of prematurity Emaciation, dehydration Diarrhoea Susceptibility to infection	Foetal stress
Meconium retention	Stoppage Ileus	Signs from birth to third day Straining, colic, lying in awkward posture Generally not off suck and in good health Abdominal tympany	Unknown

From *The Practice of Equine Stud Medicine* (Rossdale & Ricketts, 1974)

(a) Genetic, ie contributed by one or both parents.

(b) Damaged chromosomes introduced in the process of fertilisation of the egg by the sperm.

(c) Viruses and bacteria.

(d) Drugs and other substances which cross the placental barrier and enter the foetal body at some stage of its development. These are known as teratogens and include such drugs as Thalidomide and even vitamins and minerals in particularly high dosages.

The defects depend on the nature of the cause and the stage of pregnancy at which it operates. It is generally considered that the period during which the foetus is particularly vulnerable to the factors listed in (c) and (d) above is during the first forty days of pregnancy, when the foetal tissues are rapidly differentiating into the various organs and structures of the body. It is during this stage that defects such as cleft palate, 'hole in the heart', missing limbs and eyes are likely to develop; later we may expect deviations of growth to be most prominent, such as contracted forelimbs or other deviations in the skeleton.

Breeders' associations should adopt a more positive role in the investigation of congenital defects in horses, especially where they have the ideal basis for such a survey in possessing well-documented pedigrees. Measures could be taken to identify inherited defects (ie those passed from one generation to the next) and then to eliminate these by selective culling of defective genetic material.

Further research is required to determine the effect of management on abnormalities which do not have a genetic basis, and to relate the frequency of these abnormalities to management factors like feeding, administration of hormones, drugs, etc during pregnancy. At the present time we really have little, if any, understanding of the subject in horses. An inherited disease known as combined immuno-deficiency (CID) has been reported in Arabian foals and is included in *group 3*. The CID disease is rather different from haemolytic disease because, in this case, the foal cannot manufacture lymphocytes in the normal way. Lymphocytes form the basis of the immune response to infection and thus their absence results in a complete loss of resistance to infectious disease. Affected foals usually die of overwhelming infection from virus or bacteria within the first four months of life. They may have sufficient resistance in the newborn period to withstand infection because of the protective substances which they receive from the mare's colostrum.

Group 4 conditions contain that very interesting, but rare, disease haemolytic jaundice of the newborn foal. This condition results from a freak of nature whereby the foetal foal inherits from its sire a red blood cell factor (antigen) which is not present in its dam. This factor, which we may call A, is somewhat similar, though not identical, to the rhesus factor causing haemolytic disease in newborn infants. The foetal red cells contain the factor A and some of these leak into the mare's blood stream where their presence evokes an immunological response in the same way as a vaccine acts to produce antibodies when injected into the body. In this case the mare produces antibodies against factor A. These are concentrated into the colostrum along with the other protective substances (antibodies) just before foaling. Unlike the process in the human, the antibodies in the mare (mother) do not pass back across the placental barrier to the foal (infant), and thus the foetal foal is not affected *in utero,* as is the case in the infant. However, at the first suck, the foal imbibes the colostrum, and the anti-A antibodies pass, with the other protective substances, into the foal's blood stream. Here they alight on the foal's red blood cells and coat their surface, reacting with the A antigen and destroying the cells. This massive destruction of red blood cells causes a profound anaemia and jaundice.

SYMPTOMS

The reader with responsibility for the care of newborn foals is interested not so much in diagnosis of different categories of disease but in how to recognise symptoms of such a disease. The importance of early warning of disease so that remedial action may be taken in time has already been emphasised. Table H summarises the signs we should look for in newborn foals, and can be used as an aide-mémoire for those who have not, perhaps, had the experience of caring for the very young foal.

It is also helpful to take a chronological view of when signs may first appear relative to the disease process (Table I). Obviously the first concern is that the foal should breathe. Stillborn foals may be of two kinds: those which stopped living before foaling started, and those that have the potential to breathe, but fail to do so when finally delivered. Foals in the second category show some signs of life, a beating heart and perhaps a series of gasps. However, if not immediately attended to in the manner described on page 307, they make no further progress and die (ie their heart stops beating) after a minute or two.

During the first hour we may find two trends of abnormal behaviour: the one of exaggerated movements, perhaps jerky and convulsive; the

TABLE H Summary of signs to look for in newborn foals

	Normal signs	Abnormal signs
30 seconds to 1 minute	Breathing (rate about 70/min) Temperature: 37.5°C (99.5°F) Heart rate: about 50-80 beats/min	Failure to breathe or gasping indicates oxygen lack Very slow or fast rate indicates birth damage
5 minutes	Cord ruptures Shivering – lifting head and turning onto brisket Sucking } reflexes Righting	Absence of reflexes, jerking or convulsions indicates possible brain damage
	Attempts to stand and co-ordinate limbs	
30 minutes	Auditory/visual/vocal responses present Standing for the first time	
1-12 hours	Seeking the mare's udder Sucking and learning to follow the mare Breathing (rate about 35/min) Temperature: 38°C (100. 4°F) Heart rate: about 120-140 beats/min Passing urine and meconium for the first time	Unable to stand or having difficulty in getting up after 2 hours suggests weakness from infection or other causes Rapid breathing may indicate infection, haemolytic or other disease High or low temperatures suggest infection or metabolic disturbances High rate may indicate haemolytic disease, infection or maladjustment Red urine indicates haemolytic disease
	Getting up and down in one deliberate action	
12 hours	Breathing/temperature stabilised Heart rate decreased to about 80-120 beats/min at rest	
24 hours	Intestines impermeable to protein molecules	Low resistance of foal to infection if passive immunity has not been transferred by this time
48 hours to 72 hours	Increasing strength of maternal/foal bond Meconium completely voided 'Milk dung' appears	Nature of meconium important eg diarrhoea

TABLE I Time of appearance of first signs

Condition	Before 12 hours	12-24 hours	24-48 hours	48-72 hours	72-96 hours
Group 1	+	++	+++	+++	+++
Septicaemic conditions					
Group 2					
NMS	+++	++	+	+	
Prematurity, immaturity	+++				
Meconium retention	+	+++	+++	+	+
Group 3					
Patent bladder				+++	++
Congenital defects	+++	+			
CID		(Palate)			
Group 4					
Haemolytic disease			+	+++	++

From *The Practice of Equine Stud Medicine* (Rossdale & Ricketts, 1974)
    + = likelihood of signs appearing

other of an opposite nature, ie weak, where the foal is incapable of lifting itself to the standing position and where, once standing, it is too weak to approach the mare and hold the sucking position in the normal way. Convulsive or jerky movements are associated with brain damage, whereas weakness is more likely to be the result of prematurity (dysmaturity) or infection.

Convulsions and Weakness

Convulsions may occur suddenly at any age up to twenty-four hours but most frequently within three hours of foaling. Weakness is usually apparent from birth, although the observer may not appreciate the fact until the second or third hour, when the foal should be sucking with increasing strength. Foals suffering from infection may appear normal to about age two hours and then, having sucked, they show decreasing signs of strength, being incapable of holding the sucking position for any length of time. A sudden loss of suck, on the other hand, is more likely to be associated with a maladjustment-type illness. There are exceptions to this rule but in general it may be relied upon. Foals over age one week that go 'off suck' are usually suffering from diarrhoea or some other alimentary disturbance.

Breathing

The manner in which foals breathe is of interest and importance from the moment they take their first breath. Gasping at this stage is a bad sign, but once a breathing rhythm has become established the rate, depth and type of breathing movements are important indicators of respiratory function. The rate may be gauged by watching the movements of the chest and timing them over thirty seconds or a minute. In some cases of shallow breathing, it may be easier to watch the nostrils, which move in unison with the movements of the chest. We may liken breathing to the action of bellows. The same amount of air can be moved by rapid, short strokes as by slow, fully extended movements of the bellows. The quantity of air is in effect a product of depth x rate, so that twice the depth (volume) and half the rate is equivalent to half the volume and twice the rate. A young foal lying on the ground will often change its rhythm spontaneously.

Judgement of the character of breathing is a matter for vets, who can apply other tests to confirm their opinion. However, a very rapid rate (sixty chest movements per minute) maintained for several minutes in an unexerted foal is abnormal. A rapid rate after exertion or fright is to be expected, but a foal aged two to three days seen breathing very rapidly and deeply after exertion might be a suspected potential case of haemolytic jaundice. In these circumstances its heart rate would also be considerably increased (greater than 150 beats per minute) and there would be signs of jaundice on the whites of the eye and on other visible mucous membranes. Laboured breathing with widely dilated nostrils in a foal at rest is usually a sign of pneumonia or some metabolic disturbance associated with hyperacidity of the blood.

Heart Rate

The heart rate reflects the status of the vascular system. We have already seen that at birth the heart rate should be between about fifty to ninety beats per minute and any rate outside this range indicates a difficult foaling involving trauma or shortage of oxygen. The heart rate after age

Hyperflexion of the forelimb with some 'knuckling over'

Club foot seen from the front (left foot); a shoe has been placed on the tip of the foot to increase the pressure towards the heel in an attempt to correct this deformity

thirty minutes depends on the activity of the foal at the time of recording, but an excessive rate and a heart pounding at the chest wall when the foal is settled in the lying or standing position may denote haemolytic jaundice or a heart defect; both are rare occurrences whose presence must be confirmed by veterinary expertise.

Temperature

The foal's rectal temperature rises in periods of activity, such as convulsions, or in the presence of infection and pain; it falls when the foal is in coma or a state of abnormal weakness and lassitude. The body temperature may rise above or fall below normal through brain damage; therefore its measurement is not as reliable an indicator of infection as it is in older foals or in adults.

Meconium Colic Signs

Signs of colic (rolling, lying on the back or in awkward positions and excessive straining) may appear in the first few hours after birth but are more commonly associated, on the second and third days, with meconium retention. A foal should have evacuated all its meconium by the end of the third day and should be passing 'milk dung' by the fourth day. If the foal continues to strain frequently, even though meconium has apparently been cleared, we may reasonably suspect a ruptured bladder. This condition is described more fully below. The studman should note whether or not an individual is passing a good flow of urine during the first four or five days after foaling – if it is, a ruptured bladder is less likely. However, a foal that is showing signs of straining, having passed meconium, and whose abdomen is beginning to fill, should be examined by the vet as soon as these signs are noticed.

Physical Defects

Physical defects may generally be noted immediately upon delivery and as the foal is attempting to get to its feet. It should be routine practice to look for such abnormalities as a missing eye or contracted forelegs. There are cases on record where even the most experienced stud manager has reported the arrival of a foal to its owner without first making a sufficiently thorough examination to determine that one eye was missing, thus causing him to make a second and somewhat embarrassing phone call!

There are some defects which will not become apparent until later. These include:

(a) umbilical hernia, the signs of which do not appear until the foal is about four weeks old;
(b) major deficiencies in the gut causing acute signs of colic at about age 12 to 24 hours;
(c) 'hole in the heart' producing signs at age two days or older;
(d) defects such as haemophilia and combined immuno-deficiency disease diagnosed only after weeks or possibly months following birth.

THE STUDMAN'S DUTIES

Diagnosis is the responsibility of the veterinarian, and observing the first signs of disease the responsibility of studmen. The following brief description of conditions affecting the newborn is intended to further the studman's understanding of certain diseases and to place in perspective the comments already made on the symptoms and origins of these conditions. It also provides an opportunity to indicate appropriate measures of first aid and supporting therapy.

There are two situations in which the foal has difficulty in establishing a normal breathing pattern following birth:

1 The foal is subjected to oxygen shortage due to pressure on the cord before its chest has been delivered from the birth canal. It starts to breathe but cannot do so properly because of the constraint on its chest. Thus its first efforts at breathing are frustrated, and air does not enter the lungs in sufficient quantities. For a moment it stops its efforts at breathing but responds to the stimulation of falling oxygen levels and, shortly, the foal produces a series of gasps. These gasps continue for a minute or so, and if by this time the foal has become freed from the birth canal, its lungs are inflated and the oxygen concentration in the blood stream climbs dramatically. Many foals are born with 'primary apnoea' (literally: no breath). Breathing is initiated by the combined stimulus of decreasing oxygen and increasing carbon dioxide levels (associated with apnoea) and the external stimuli of cold and friction.

2 In this example, the foal is still in the birth canal when the series of gasps described above occurs. The lungs, in this case, do not fill with air, and after a minute or two of gasping the foal stops breathing again, going

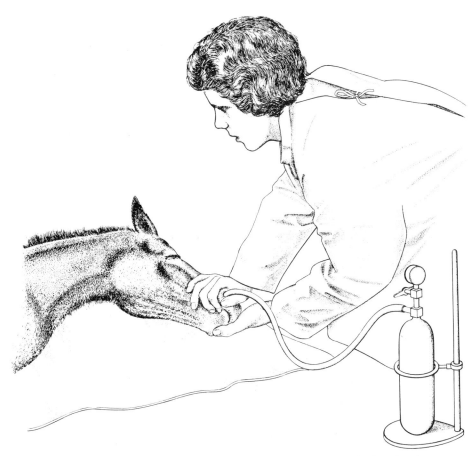

Fig 33 The method of artificial respiration using an oxygen cylinder: for the sake of illustration the tube is entering the upper nostril, showing how one hand is placed round the nostril closing it over the tube and the other hand is used to close the under nostril while the operator watches the foal's chest for signs of movement as the oxygen flows into the lungs. In practice, it is preferable to put the tube into the lower nostril and to release the upper nostril to allow the foal to 'breathe out' once the lungs have been sufficiently inflated

into 'secondary apnoea'. It will not recover from this phase until the lungs are artificially ventilated and other measures taken.

It is not possible for those present at foaling to distinguish primary apnoea (when the foal will start breathing spontaneously after gasping) from secondary apnoea (requiring artificial aids). Of course, if one waits to determine in which state a particular individual is born, the answer becomes clear over the course of five minutes, but if the foal happens to be in secondary apnoea it will be dead before the distinction can be made. Secondary apnoea must be treated by artificial inflation of the lungs with oxygen and then the administration of alkali directly into the blood stream, to counter the rising acidity associated with asphyxia. It is the acidity of the blood that is fatal, causing the heart to stop beating. Inflating the lungs achieves an increased concentration of oxygen but may not, on its own, save the foal if the acidity of the blood is not counteracted at the same time.

Methods of Artificial Respiration

The lungs may be inflated by mouth-to-nostril artificial respiration or, preferably, by using an oxygen cylinder attached to a rubber tube which is inserted into the lower nostril (fig 33). The operator should kneel at the head of the outstretched foal, facing from the front. From here one hand is

Fig 33a Mouth-to-nostril respiration: the right hand is placed around the upper nostril to which the mouth is applied and the left hand is placed over the under nostril to close it when the operator is blowing into the nostril. The method may be used with the foal lying on its left or right side and it may be an advantage for the operator to kneel at right angles, ie parallel to the foal's forelegs

placed under the foal's muzzle and the rubber tube inserted to a distance of about three inches (six centimetres) into the lower nasal passage. The lower nostril is occluded by pressing the rim around the tube, and oxygen is run from the cylinder at a flow rate of about 10 litres per minute. The upper airway is closed with the other hand, and the operator watches for movement of the chest wall as the lungs inflate. It is most important not to overinflate and it is sufficient to proceed only until the barest movement of the chest is observed. The pressure should then be released by removing the hand from the upper nostril. The process should be repeated at intervals of about ten seconds until a spontaneous breathing rhythm is established and the foal is able to continue breathing on its own. The period before it will breathe unassisted depends to some extent on the amount of asphyxia it has suffered. If the brain has been damaged by the lack of oxygen the foal may suffer from convulsions. If a foal has to be assisted in the manner described it is prudent to call for urgent veterinary assistance. Alkali should be administered as soon as possible in these circumstances, but this procedure is outside the scope of first aid unless the attendant has been schooled in making intravenous injections. The amount required to combat acidity is about 4fl oz (100ml) of a five per cent sterile solution of sodium bicarbonate. This may be administered rapidly by means of a syringe or through a gravity-feed system.

Artificial respiration may also be administered by means of applying the mouth to the uppermost nostril and closing the lower nostril, proceeding in the same manner as described above (fig 33a).

Signs of Infection

Signs of infection include sleepiness, lethargy, gradual loss of the suck reflex, and an inability to hold the position of sucking. Foals affected in the first twelve hours have probably been infected during pregnancy and born with the microbes already active in their tissues. Post-natal routes of microbial entry include the mouth, airways and umbilicus. We have already noted the harmful effects of dust in the foaling box which may help to convey microbes into the lungs. Precautions to reduce the likelihood of microbes entering the umbilicus include the provision of clean bedding and hygienic conditions, especially when handling the umbilicus. Normally the cord breaks at a natural point about 1½in (3–4cm) from the abdomen, and the vessels shrink so that entry to the foal's abdomen is closed and sealed. Applying iodine or antiseptic and antibiotic powder to the stump is not necessary in most cases, because nature is more effective.

However, a thick stump or one resulting from an abnormal rupture may require antiseptic dressings.

The ability of microbes to cause disease depends on their nature and virulence and on the resistance of the foal. We have already discussed the manner in which the foal attains its immune status through the transfer of protective substances in the colostrum. The susceptibility of any particular individual to infection is therefore largely related to the efficiency of the passive transfer of immunity. To prevent infection is partly a matter of ensuring an adequate supply of protective substances and partly of avoiding contact with virulent microbes.

If a mare has 'run milk' or there is reason to believe that the foal is unlikely to receive its share of good-quality colostrum, some other source must be found. Many studfarms nowadays keep samples of colostrum collected in each breeding season from mares that have an abundance. It is important, when collecting samples, to avoid depriving the donor's foal. Only small quantities (2fl oz (50ml)) should be taken in each instance. Larger quantities can be collected from those mares whose foals have died during or shortly after birth. Colostrum from mares with haemolytic antibodies should not be used.

The samples are stored in a deep freeze at 0°F (–20°C). Of this colostrum, 10fl oz (200–300ml) should be administered by means of an artificial teat to a foal that has a strong suck reflex or through a stomach tube to one that cannot or will not suck from a bottle. The colostrum should be fed at blood temperature from clean utensils before age twelve hours. It should be given before any other feed, and although the dose may be divided it is preferable to give the maximum amount in one feed because the ability of the foal to absorb the protective substances decreases once protein has entered the intestines. Thus a feed of non-colostrum, prior to administering colostrum, actually reduces the foal's ability to absorb the protective substances.

It is impossible to enhance the immune status of the individual by administering colostrum after age about twelve hours, because the intestines become impermeable to the antibodies. Blood may be collected from a donor horse, the plasma separated and infused slowly into the foal's blood stream through a needle or catheter inserted into one of the large veins. This technique is somewhat laborious, and the horsebreeding industry requires a reliable substitute such as a freeze-dried protein product for subcutaneous injection. However, the commercial potential of such a product is apparently insufficient to sustain the interest of pharmaceutical firms in this area.

The treatment of infectious conditions is a matter of veterinary judgement, based on an assessment of the case and the probable nature of the infecting microbe. In general, frequent large doses of antibiotics supplemented by the administration of plasma and other supportive measures provide good results. However, it cannot be too highly emphasised that successful treatment depends on applying measures as early as possible after the start of infection; and thus on the ability of studmen to notice and interpret signs of disease in its earliest stage.

The Approach to Conditions of Prematurity, Dysmaturity and Immaturity

Apart from being undersized, premature foals may have one or more of the following characteristics: a silky fine-haired coat, deeply red tongue, inability to get up for several hours after foaling, difficulty in reaching the udder or of holding the sucking position, a tendency to develop 'bedsores', and a lack of fat giving an undernourished (emaciated) appearance. The premature foal is usually more susceptible to infectious diseases and diarrhoea than are foals born at full term. Premature foals may survive and thrive, although some individuals subsequently develop problems of bone development and may remain undersized.

Dysmature foals have been deprived of full placental function. The extent of their deprivation depends on the area of placenta that has been damaged and the stage of pregnancy at which the damage occurs. The foal will be undersized if the damage occurred relatively early in pregnancy (eg before the ninth month) but it is likely to be full grown but undernourished (emaciated) if the damage occurred late in pregnancy (eg during the last month).

Twins are a special example of dysmaturity. We have seen on page 272 how the competition between two placentae damages both and limits their area of attachment to the wall of the uterus. Thus each member of twins is deprived to a lesser or greater extent, depending on the area of attachment and its functional integrity. Twins that are born alive are usually small and exhibit typical signs of prematurity and dysmaturity, either surviving with assistance or deteriorating and eventually succumbing to an inability to adjust to the extra-uterine environment.

Dealing with Cases of Neonatal Maladjustment Syndrome (NMS)

This term describes foals suffering from convulsions ('barkers'), sudden loss of the suck reflex and the ability to recognise and follow the mare ('wanderers' or 'dummies'). Symptoms usually appear within twenty-four hours of birth, and most often within the first three hours. Severe and mild forms occur; the two following case examples illustrate the extremes:

1 A foal is born with an easy delivery and appears to be normal until age about thirty minutes. At this time it is struggling to get to its feet. Suddenly, it starts to develop jerky movements of the neck and forelimbs. It stays in a dog-sitting position with its head and front part of its body jerking spasmodically. Soon, it emits loud sounds similar to a dog's bark – hence the term, originally coined for this condition, barkers. This phase is followed by strenuous muscular activity in which the foal makes frenzied galloping movements on its side; or if affected in a standing position, it may gallop into a wall. The body temperature rises with the convulsions, and the foal sweats profusely. Some foals may be affected after having sucked from the mare, but once symptoms start, the suck reflex is completely abolished. This sign helps to distinguish the condition from infections where the suck reflex is usually lost gradually.

The convulsive phase may last several hours and recur after remissions lasting hours or, even, days. The convulsions are usually followed by periods of coma. Foals recover if they do not die during the convulsive phase, provided they receive adequate support and therapy. However, it may take weeks for the powers of normal behaviour to be fully regained. Recovery of its faculties for getting up, moving about, sucking and following the mare usually re-appear in this order.

2 The milder forms of the condition do not involve convulsions, but the foal displays an apparent unawareness of its surroundings and, in particular, of the position of the udder, the teats and the presence of milk. The foal may stand in the same position (dummy) or walk ceaselessly round the box, perhaps bumping into walls or other objects (wanderer). These cases may recover after a matter of hours or take several days before they eventually are able to behave normally. In the interim period it is, of course, essential to maintain their strength and vitality by catering for their nutritional requirements under veterinary supervision.

Symptoms of NMS are related to brain damage, probably caused by oxygen shortage during the birth process and/or by the enormous pressures to which the foal is subjected at this time from the mare's expulsive efforts. Small amounts of haemorrhage or oedema (leakage of water from the blood vessels) may cause pressure on vital areas of the brain, accounting for the symptoms and the variable course of the condition. Treatment depends almost entirely on supportive measures combined with the use of drugs to control the convulsions and reduce the brain damage. The supportive measures are described later (page 316); good nursing by studfarm staff is an essential ingredient to successful treatment. There is no known method of prevention.

Recognising Cases of Meconium Colic and Ruptured Bladder

The foal may suffer signs of colic (rolling, lying on the back or in awkward attitudes, such as with the head tucked in between the forelegs). The symptoms are caused by a sensation of pain stemming from pockets of gas distending the intestines. An affected foal may go off suck for short periods and will strain and squat with its tail raised in efforts to pass the obstructing meconium from the rectum. Treatment should be under veterinary supervision.

Meconium colic must be distinguished from the condition of ruptured bladder, and successful treatment depends on early diagnosis. Symptoms may be similar to meconium colic, especially those of straining to pass faeces or urine. On the second or third day, the abdomen starts to swell with urine. The distension has to be distinguished from the gas of meconium colic.

The cause of the tear, usually in the roof of the bladder wall, may be pressure during birth on a bladder distended with urine. The urachus, ie the duct conveying urine from the bladder into the allantois, may be blocked prior to birth, thus preventing the escape of urine from the bladder. However, some cases are due to failure of the bladder wall to close completely during embryological development. In all cases, the peritoneal lining over the bladder may maintain the integrity of the wall until the bladder expands after birth, due to increased urine forming as the foal swallows quantities of milk. At some stage, the tension rises so that the weakened wall is breached. Urine then escapes into the abdominal cavity instead of being passed through the urethra.

The ability of the foal to pass urine in the normal manner depends, to a large extent, on the size of the aperture in the bladder wall. A small, pinhead-size hole causes a leak which does not prevent the foal from urinating normally, but allows urine to escape slowly into the abdomen. We may expect symptoms (enlargement of the abdomen) to develop after the third day. A large tear prevents the foal from urinating in the normal manner, and it may crouch frequently, passing only very small amounts. Large quantities of urine accumulate in the abdomen, and signs appear on the second or third day. The urine in the abdomen does not cause irritation but by its volume presses on the diaphragm, eventually stopping the breathing movements. The foal dies from asphyxia unless the condition is relieved by surgery.

The Significance of 'Pervious Urachus'

The urachal duct, which runs from the bladder to the allantoic cavity in the placenta, is broken when the cord ruptures at birth and becomes sealed along with the blood vessels. In some cases, however, the duct remains open and urine drips from the navel for some hours or days after foaling. The condition may be recognised by the wet patch that forms around the navel as the coat becomes soiled with urine. There is a risk of infection developing in the surrounding tissues, and veterinary advice should be sought. The condition may respond to simple treatment such as chemical stimulation with one-per-cent formol saline, or may require surgical treatment and injections of antibiotics.

Explaining Haemolytic (Jaundice) Disease

Even though this disease is uncommon, studmen should be watchful for signs of the condition, on the second and third days after foaling. Signs include a rapid breathing rate at rest, becoming laboured after the slightest exertion. Heart rate is increased at rest and pounds against the chest on any exertion. A pulse wave may be seen travelling up and down the jugular furrow of the neck. The mucous membranes are yellow and the urine red. Immediate and urgent veterinary attention is required as soon as signs are observed. The destruction of red blood cells is progressive and if the level reduces below $2 \times 10^{12}/l$ (the normal is $7–9 \times 10^{12}/l$), the oxygen-carrying capacity of the blood may fall below the critical level for life.

Treatment consists of transfusing red blood cells to maintain the necessary level circulating in the blood stream. The optimal source of these is the mare, but they must be washed free of plasma by suitable laboratory techniques because the dam's blood plasma contains the antibodies that are the cause of the condition. The mare's blood may be tested before foaling to identify the presence of these antibodies; if present, the disease can be prevented by muzzling the foal for twenty-four hours before it is allowed to suck from its dam. During this period it should be fed colostrum from another mare, followed by a suitable milk substitute. The mare's udder should be milked at regular intervals to eliminate the colostrum and stimulate the gland to produce milk in preparation for when the foal can be allowed to suck normally. In this way the foal is prevented from taking colostral antibody; any that is swallowed after age twenty-four hours will not be absorbed from the gut into the blood stream. The colostrum stripped from the affected mare should be discarded. It is not suitable for use as donor colostrum to other foals.

Signs of Congenital Defects

Symptoms depend on the site and nature of the abnormality. For example, a cleft palate (the failure of the two sides of the roof of the mouth to close in the mid-line) causes the foal to regurgitate milk down its nostrils after sucking. Defects of the heart valves and wall produce symptoms according to their position and size, including lassitude, fainting and a rapid heart beat. Defects in the alimentary tract give rise to symptoms of chronic loss of weight and colic. Most other defects are external and can be diagnosed by an examination of the individual.

The Importance of Nursing

The newborn foal must meet the challenges of its strange environment and adjust to independent existence. However, there are occasions when the mechanisms of adjustment are seriously impaired, eg when the brain or heart are damaged during birth, and in states of prematurity and dysmaturity. Treatment in these cases consists very largely of supporting life until the damage has been repaired and the mechanisms of adjustment have had time to take effect. Good nursing is essential for as long as it is necessary for the foal to establish its biological equilibrium. Thus if a foal has no suck reflex it must be maintained by artificial means of nourishment; if it cannot get to its feet it must be helped and if it is unable to maintain its body temperature it must be kept warm. Nursing in this context is therefore largely a matter of judging the foal's behavioural and metabolic deficiencies and making these good in the most appropriate manner – for a matter of hours, days or, even, weeks.

Good nursing by studmen is essential. Vets can administer sedatives, anticonvulsants and other drugs to counter the symptoms of brain and organ damage, but their treatment is of little consequence if proper nursing is not provided. Intelligent observation is required, followed by remedial measures diligently applied with gentleness and understanding of the foal's plight. For example, a foal suffering from brain damage cannot easily co-ordinate its movements to turn from the position of lying

Fig 34 Method of restraint: foal lying quietly

Fig 34a Restraint during mild struggling: note the operator tilting head to avoid upper hind limb and turning foal's head upward while grasping uppermost leg

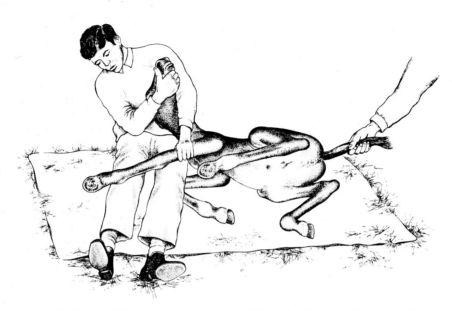

Fig 34b During violent struggling or convulsions, the foal's tail is grasped by the second attendant to prevent the foal's body moving at right angles to holder

Intensive care foal bed, designed by Dr Tim Cudd, Lexington, USA

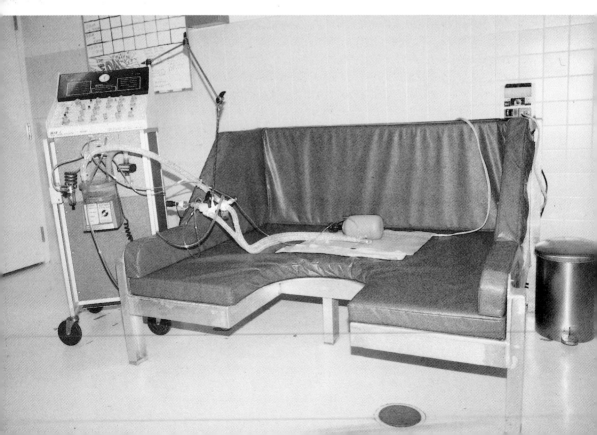

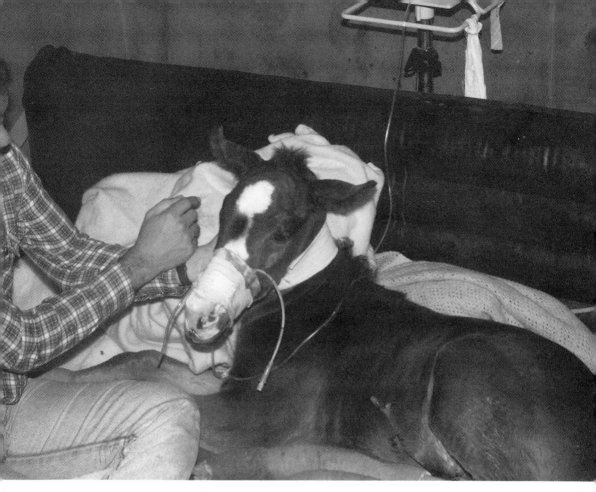

Foal undergoing intensive care on an inflatable foal bed in Newmarket

outstretched to sitting on its brisket. It may struggle violently in its attempts to achieve the change in posture. The more it struggles and fails to achieve its objective the greater will be its efforts to do so. It may consequently resort to wild thrashing movements with its limbs and hit its head violently against the ground. If the foal is helped into the upright position so that it rests on its brisket this may, in itself, calm the foal, and having achieved its immediate objective it may cease to struggle. The same sequence of events might be found in cases where the foal is struggling to get to its feet – help at the opportune moment resulting in a calming effect when the foal achieves the standing position. The drive behind the foal's behaviour stems from the desire to get to its feet and to suck. Thus assisting the foal to fill its stomach may in itself have a sedating influence.

Struggling is an energy-consuming occupation – just as much as, if not more so than, galloping. The energy is produced by burning sugars and protein stored in the tissues, and is brought to the muscles in the blood stream. Blood circulates due to the pumping power of the heart, and this is normally sufficient to maintain the nourishment and oxygen requirements of the muscles at all times. However, in the newborn foal, the energy reserves are soon depleted. The pumping power of the heart may be reduced by damage sustained during birth and by insufficient stores of energy-producing substances to sustain the power of the heart muscle. Further, the heart muscle may be damaged by chemical and toxic substances produced during asphyxia. The heart may, therefore, become incapable of meeting the demands imposed on it by the struggling foal. Heart failure is a common cause of death in newborn foal disease, and it is in this light that we must do all we can to prevent the foal from struggling unnecessarily or exerting itself during illness. Noise and rough handling should be avoided at all costs; both are likely to stimulate the foal to activity during a period when rest and calm are of utmost importance.

Holding the foal's legs together and other methods of forceful restraint are to be avoided because these accentuate the demands on the heart, lungs and nervous system, which may already be taxed to the full. It is quite possible to tip the balance between successful adaptation and malfunction towards a fatal outcome, merely by unnecessary restraint.

An attendant should be present at all times when the foal is unable to right itself or get to its feet unaided. The constant presence of an attendant may have a soothing influence on the foal in periods of nervous derangement. The position of nursing is illustrated in fig 34 (see also page 317). A programme of constant attention imposes considerable strain on stud personnel, especially as the conditions requiring attention often occur during the busy period of the stud breeding season. Nevertheless, it cannot be overemphasised that some cases depend for their survival on maximal effort in attending and nursing the foal in an appropriate manner. A successful conclusion to a case that has continued for many days is highly gratifying to those responsible. Owners, who are not usually present on these occasions, should recognise the skill and endurance of those responsible for nursing procedures.

The optimal surface on which to lay a recumbent foal has not yet been devised. The essential needs are for a soft undersurface against which the foal will not bruise its prominent parts, such as the eyes, hips and shoulders. But it must be sufficiently porous to allow adequate drainage of urine to prevent 'scalding of the coat'. Bedsores occur in foals that have been lying on their sides for many hours. Talcum powder helps if used in liberal quantities on the prominent parts; the foal should also be turned from one side to the other at hourly intervals to allow evaporation from the moisture-laden coat.

Foals that find difficulty in getting up and/or holding the sucking position should be helped. Common sense applied to each individual case is usually sufficient to avoid unnecessary struggling and fighting against the foal's natural instincts. It is important to interpret the desire of the individual and to discover whether it resents or accepts any interference. There is no truer saying than 'you can lead a horse to water but you can't make it drink' when describing human assistance in helping foals to suck from the udder.

Keeping Warm

The balance of body heat is the sum of heat produced by the muscles and other tissues minus the heat lost through evaporation, conduction, convection and radiation. A foal in coma has reduced output of heat, normally supplied by muscular activity and by shivering; in coma there is no such activity. It is therefore necessary to reduce heat loss to the minimum by

Helping a premature foal to maintain the sucking position

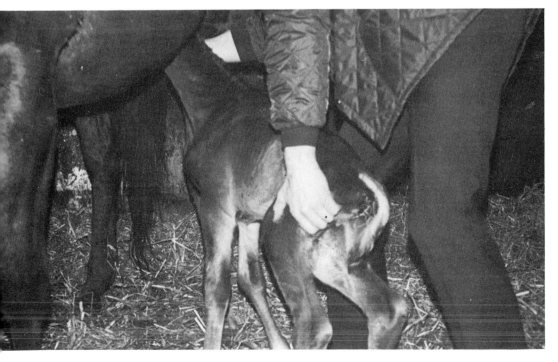

raising the surrounding air temperature. To cover the foal with a rug or several blankets may not be sufficient in maintaining its body temperature to near to 100.5°F (38.5°C) in severely affected foals. Electric blankets with a washable moisture-proof surface may be used with good effect if placed beneath a blanket covering the foal. The rectal temperature should be recorded hourly, and applied measures varied according to the results to maintain the desired range in the region of 100.5°F (38.5°C).

Feeding

Feeding and the administration of fluids by mouth or intravenously lie within the province of the veterinary surgeon attending the case. Consequently it is not appropriate to advise readers on the quantity, quality and frequency of diets administered to sick newborn foals. The same approach applies to drug therapy and other veterinary medical support therapy. The reader will appreciate that, in any case, these measures depend on recent knowledge and past experience, both of which are the responsibility of the vet and not of the owner or attendants.

DISEASES OF THE OLDER FOAL AND WEANLING

The diseases of this period of life are, with a few exceptions, comparable to diseases encountered in adults. The studman is concerned with recognising the signs of these diseases, and in the present context this seems the most appropriate route by which to approach and describe the conditions.

Diarrhoea

Diarrhoea (scouring) is a symptom of several different conditions. Its appearance may be of little or of considerable significance, depending on its cause. The incidence of diarrhoea in foals up to five months of age is much higher than in any other category of 'patient'. Diarrhoea may be the result of managerial, nutritional, bacterial, viral, fungal or parasitic factors. Diagnosis is the responsibility of the vet, but the studman must be mindful of the possible causes in relation to the spread of the condition to other foals. If a foal is scouring, the following should be noted:

(a) Is the dam in heat? The foaling heat in particular may cause scouring in the foal. The diarrhoea is usually mild but it may develop into a more severe form in the presence of other factors, such as infection.

(b) The rectal temperature should be recorded, taking the condition of the anal ring and rectum into account. If the ring is dilated and there is air in the rectum the recorded temperature may be false because the thermometer will not be in direct contact with the rectal mucous membrane.

(c) A sample of faeces should be collected into a clean, preferably sterile, bottle for analysis by the veterinarian. The diarrhoea may be green, brown or grey coloured; of an offensive odour or odourless; have a consistency of water, porridge or be pasty; and may be tinged with blood.

(d) The appearance of the tongue may be moist and normal or dry and furry.

(e) The individual may be dull and depressed or bright and vigorous, off suck, feeding normally or drinking large quantities of water.

The course of the condition varies from cases that 'purge' for a few hours or those that are affected for days, even weeks. Studmen must become experienced in assessing whether or not veterinary attention is urgently required. A dull, depressed individual which is off suck and drinking copiously at the water manger requires immediate treatment. A foal that is strong and otherwise apparently quite healthy may be left for hours, or until the following day, after the appearance of first signs before professional assistance is called.

An affected foal should, however, be isolated as far as possible from other foals of a similar age and susceptibility. Isolation in a loose box is the most convenient method, but it should be appreciated that some agents causing diarrhoea, such as rotavirus, may accumulate on the bedding and even in the manger, thus re-infecting the foal and causing the condition to worsen. Scouring foals should therefore have their bedding replaced frequently, and areas of faecal contamination disinfected. If practical, it may be helpful to place the foal in a second loose box while the first is cleaned.

Rotavirus

Rotavirus affects foals up to five months of age. Its presence can be confirmed from a faecal sample. The most common clinical manifestation is profuse diarrhoea but other signs include depression, anorexia, mild colic, increased gut sounds and fever. The period of diarrhoea usually lasts up to two weeks. Severe cases may result in dehydration and metabolic acidosis. The virus damages the intestinal tract lining which can result in a failure to absorb nutrients satisfactorily. The condition can be treated with fluids, and antibiotics may be necessary if secondary bacterial infection occurs. To date, no vaccine has been developed against foal rotavirus, although cows are vaccinated before calving so that antibodies are present in the milk. The cattle vaccines have been used in horses but may cause a severe local reaction.

Summer Pneumonia

This is a condition of young foals (one to five months old) characterised by multiple small abscesses in the lungs. Symptoms are those of pneumonia, but the disease is insidious in onset, and damage to the lung may be well advanced before signs are noticed. Symptoms include a rapid and laboured breathing rate, high persistent fever lasting for days or weeks, coughing, nasal discharge, loss of condition. It has a fatal outcome, although some cases recover if given intensive treatment with antibacterial compounds, Erythromycin and Rifampkin, sufficiently early in the course of the disease.

Epiphysitis

Inflammation of the growth plates (see page 227) usually occurs in the period of greatest activity, ie some time before growth of the part ceases and the plate closes. Thus we may expect to observe signs at the lower end of the cannon bones when the foal is about six months old, and in the lower end of the radius (forearm) at age eighteen months. Painful swellings sometimes develop on the inside of the growth plate. Affected individuals may be lame during the acute phase of the condition and be subject to unsoundness when broken to tack.

There are a number of different causes of epiphysitis, but excess pressure due to poor conformation and uneven weightbearing on the limb is the most common cause. Overweight, malnutrition, diets containing excess phosphorus and too little calcium, vitamin D deficiency and diets rich in protein may be directly linked or act as predisposing causes. There may be some hereditary susceptibility in some cases.

Veterinary advice should be taken to diagnose the cause and provide the most favourable means of treatment. The calcium/phosphorus ratio of the diet should be about 2:1, and the protein content of the diet less than ten per cent of dry matter for affected individuals. Bran should not be fed; calcium carbonate or bone flour ($^1/_3$–1oz, 10–30 grams, daily) should be added to the diet.

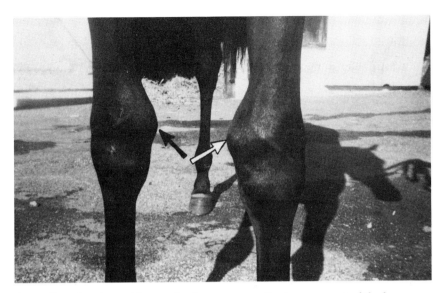

A yearling showing evidence of enlargement on the inner aspect of the lower end of the radius (forearm) denoting epiphysitis

Contracted Fore and Hind Limbs (Hyperflexion)

Many foals are born with straight limbs or 'knuckling over'. The condition may improve with age, as the foal takes exercise and pressure remoulds the limbs by its action on ligaments, joints, tendons and bone. In other cases the condition gets worse, or sometimes it develops, suddenly or gradually, when the individual reaches an age of two months or more. The cause of the condition is unknown, although deficiencies of calcium, phosphorus and vitamins A or D have been suggested. Inherited suscepti-bilities may well be involved in combination with overfeeding, exercise on hard ground or infection.

Affected individuals are usually treated by attempting to correct the deviations by opposing pressures, such as by lowering the heels and/or applying a shoe to the toe of the affected limb. Plaster casts and corrective boots may be applied to very young foals. However, in all cases, treatment can be very unrewarding. By lowering the heels and placing a tip on the foot we may correct a tendency to knuckle over at the fetlock joint but, at the same time, produce strain on the knee joint which causes the front of the limb to become concave, ie 'back at the knee'.

In more recent times severing the sub-carpal check ligament (ie the ligament attached to the back of the knee at one end and the deep flexor tendon at the other) produces immediate relief in many cases. The operation is performed under general anaesthesia and allows the toe to come forward and the heels to be lowered merely by releasing the tension on the deep flexor tendon which is inserted into the toe (pedal bone). The disadvantage of the operation is that it may leave a small blemish at the site of the surgery. However, the great majority of foals develop normally and achieve racing soundness. The administration of phenylbutazone may aid recovery and, also, be used in mild cases to effect recovery without the need for surgery. However, each individual case requires veterinary judgement as to the optimal approach.

Joint-ill (Infective Arthritis)

Hot, painful swelling of one or more joints with associated fever and lameness are the features of this condition, affecting foals aged from a few days to four or five months. The disease is essentially an infection of the cartilage and bone adjacent to the joint, but the joint membranes and surfaces may also become inflamed and ulcerated.

A variety of microbes may be found in the joint or in the abscesses beneath the joint surface, most commonly *Streptococci, Staphylococci, Salmonella* and *Corynebacterium equi.* The microbes enter through the umbilicus or the alimentary tract and circulate in the blood, becoming localised in the joints, where they find the conditions conducive to their growth. Early diagnosis and treatment are essential, and veterinary advice should be sought if a foal is lame. The foal should be confined to a loose box and its rectal temperature recorded. Lameness without fever is probably not due to infection, but it is important to record the temperature twice daily and to regard any rise above 101°F (39.5°C) as significant.

Infections of the Breathing System

Foals and yearlings are particularly prone to infections of the respiratory tract. Symptoms include catarrhal exudates from the nostrils ('snotty nose'), coughs, bronchitis and pneumonia. The infections are caused by virus, often followed by a secondary bacterial infection such as of *Streptococci* or *Staphylococci.*

The snotty nose condition, in the absence of more serious signs of fever or pneumonia, is not regarded as particularly serious. In fact the condition may be beneficial by promoting the individual's immunity. However, the condition is unsightly and a nuisance when foals or yearlings are being prepared for sale. The catarrhal discharge may be persistent over many days or even weeks, recurring perhaps at intervals. The discharge may be primarily from the sinuses of the head or from the lungs. The condition should be distinguished from strangles, which is a specific infection of the lymphatic and upper respiratory tract caused by a microbe known as *Streptococcus equi*. This microbe should be distinguished from the more common *Streptococci* which are found on most damaged surfaces in horses. In cases of strangles, fever and swollen glands of the head and throat are present. The differential diagnosis is the responsibility of the vet, who should always be called for advice if a foal or yearling is found to be suffering from fever, enlarged glands and nasal discharge.

Umbilical Hernia

A swelling appearing at the umbilicus when the foal is about six weeks old, steadily increasing in size, may be an umbilical hernia or an abscess. If we gently squeeze the swelling the hernia will reduce, whereas an abscess remains as a firm non-compressible swelling. Hernias are

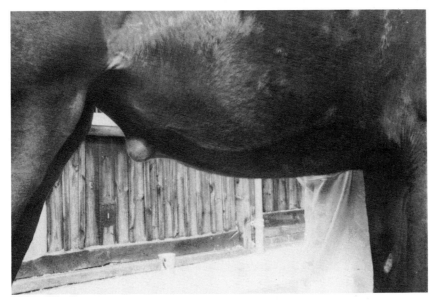

Umbilical hernia: a soft swelling appears at the navel at about two months of age; this swelling can be distinguished from an abscess by gently pinching the swelling. If it is a hernia it will reduce

unsightly but usually disappear by the time that the affected individual is a year or two years old. However, the contents may become strangulated as the ring through which the intestines protrude from the abdomen closes. This causes the individual to go off its feed, appear dull and show evidence of colic. Surgery is essential in such cases, and the veterinarian should be called urgently for advice. It is nowadays quite customary to eliminate the hernia surgically when the foal is about three months old. This removes the blemish and the risk of the hernia becoming strangulated.

Tetanus (Lockjaw)

This disease gets its names from the typical spasms which affect the muscles of the limbs, neck and jaws, and from the causal microbe *Clostridium tetanus*. The microbe lives in soil and faeces, contaminates wounds and enters the body through broken skin surfaces. The newborn foal is particularly vulnerable to the entry of this microbe through the navel, where ideal conditions exist for growth within an internal or external abscess at the umbilical ring. The microbe is destroyed by oxygen and can breed only in pus or under scabs where oxygen is

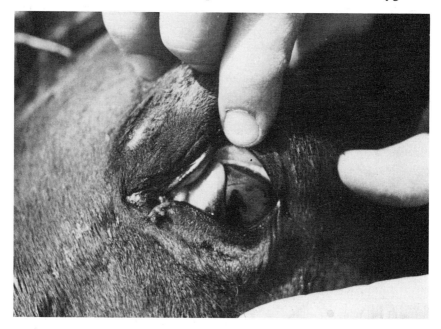

The third eyelid can be seen crossing the eyeball in a spasm due to tetanus

excluded. It produces a toxin that travels along the nerve trunks and affects their functions, causing the characteristic muscular spasms.

The first signs are stiffness and reluctance to move. Spasms develop rapidly, increasing in severity until the powerful muscular contractions affect all parts of the body, each move triggered off by noise or handling. The ears are pricked forward, the head outstretched and the base of the tail raised to the horizontal. If the affected individual lies down or falls over on its side it may be unable to get to its feet again. There is difficulty in swallowing and passing faeces. Most cases end fatally, although with modern treatment less severely affected cases may recover after some weeks.

Lockjaw is usually more prevalent on one farm or premises than another; and it may even be restricted to one field within a farm.

Foals are most susceptible to tetanus in the first six weeks of life. Protection may be achieved by vaccinating the mare or injecting the foal with antitoxin at birth and again when it is about thirty days old. For the protection of individuals, the vaccine called toxoid is injected once, followed by a booster at four weeks, another at twelve months and then annually. Individuals over the age of three months can thus be protected against the disease. Foals under three months receive protection from their dams through antibodies in the colostrum, provided the mare has herself been immunised. A booster dose of toxoid within one month of foaling increases the level of antibody in the colostrum.

Tetanus antitoxin serum is produced by immunising horses and then bleeding them. The antibodies contained in the blood serum collected from these horses are concentrated to provide doses that may be injected into individuals to provide temporary protection. A dose of antiserum may be given when a horse has sustained a wound and it will protect against the disease for about thirty days. After this period the protection wanes, and the horse is again susceptible to infection.

Equine Parasites

The menace of redworm and other parasites has long threatened the health and development of horses, and is especially damaging to those used for performance, such as Thoroughbreds. It is these breeds that are also at greatest risk because of the tendency for over-concentration of stock over many years on pastures traditionally used for the purpose of breeding. One of the most significant advances in recent years has been the intro-duction of the drug Ivermectin. This has the distinct advantage over other

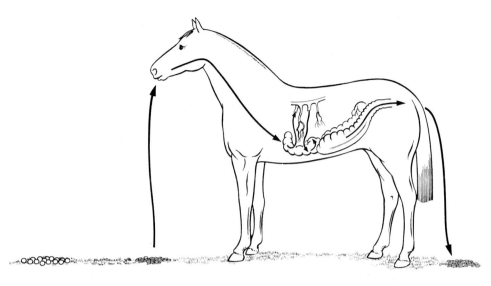

Fig 35 Life cycle of Strongylus vulgaris *illustrating that larvae are grazed from the pasture, enter the small intestine and travel in the blood stream to the aorta and thence back to the colon where they become adult and lay eggs, thus completing the life cycle*

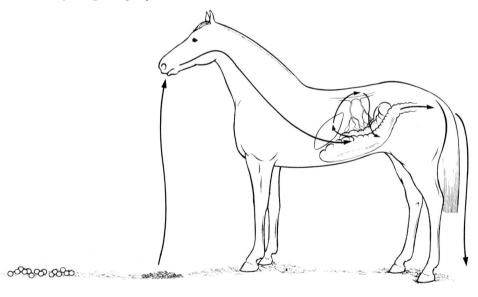

Fig 35a Life cycle of Strongylus edentatus *which has a similar life cycle to S.* vulgaris *(fig 35) except that the larvae enter the peritoneum and not the blood vessels. Having returned from the peritoneum they reach the colon and lay eggs which are passed in the faeces*

compounds of killing parasites not only in the gut but in the tissues and blood stream where the young forms of the parasite reside for varying periods. Another advantage of the drug is that it kills most classes of parasites including red and white worms and bot and warble fly larvae. Therefore, young stock especially should receive regular doses of this particular drug, although it may be given alternately with other, less expensive, compounds. It is important to kill the adult forms in the gut as well as the larvae during their development in the body tissues themselves. There are a number of parasites which live inside horses. Some of these are harmless and others highly dangerous. The most damage is caused by that family of parasites known as redworm or strongyles. There are a number of differing genera and species of this family, the most harmful of which are *Strongylus vulgaris* and *S. edentatus*. The life cycles are illustrated in figs 35 and 35a. It will be seen that *S. vulgaris* lives for part of its life in the blood vessels of the small intestine; the larvae of S. *edentatus* live in nodules in the peritoneum. Thus, apart from the damage done by all gut parasites, the life cycle of these two species takes them into the tissues of the body, posing added risks.

Redworm infestation is insidious in the sense that symptoms are usually ill defined, ie loss of weight or failure to thrive, and the effects are long term. More acute symptoms include colic, resulting from damage to the blood vessels supplying the gut wall, causing it to become inflamed and painful. Redworm infection can cause death through massive damage to the gut wall leading to rupture. In young animals the larvae of *S. vulgaris* may weaken the wall of major blood vessels, causing their rupture and sudden death through haemorrhage.

Redworm control is essential to good studfarm management. The following principles should be observed.

1 Infection is picked up during grazing, and therefore the number of eggs deposited on pasture in dung must be restricted to a minimum.
2 Regular administration of anthelmintic drugs at, say, monthly intervals reduces the number of parasites in the gut and therefore the number of eggs passed in the faeces.
3 The number of eggs per gram of faeces may be measured in a laboratory; if this test is performed at, say, three-monthly intervals, prior to treatment, it provides an indication of the efficacy of the worm control programme.
4 It may be prudent to alternate between different drugs administered each

month so as to avoid the possibility of the parasites becoming drug-resistant.

5 Pastures should be rested for three months or more to reduce further the number of larvae available for re-infection. A substantial proportion of larvae will die during this period, provided they are exposed to sunlight and conditions are such as to exhaust their energy.

Larvae cannot feed on pasture, and depend for their survival on being picked up by a grazing horse. If this does not happen before their energy reserves are consumed they die. Alternate bright and dull conditions cause the larvae to move up and down the blades of grass, thus reducing their life span; and bright sunlight dries them. Thus there is a steady decrease in numbers with every week that a pasture is rested, although some larvae may remain after many months, and even ploughing and resting ground for a year will not eliminate entirely the larvae present.

Picking up dungs on a daily basis, grazing sheep or cattle on the pasture and harrowing the ground are measures that are recommended to reduce further the parasite burden of pastures. The burden is proportional to the number of horses per acre grazing the pasture and the time they spend on it. Roughly four times the amount of dung will be deposited on a pasture grazed by two horses per acre day and night than by one horse per acre for only twelve hours per day.

Special attention should be paid to pasture grazed by foals, yearlings and two-year-olds, because young horses are more susceptible to damage by the redworm larvae than are older horses. Horses obtain some immunity to the effects of infection by exposure to larval migration and the presence of parasites in their gut. Perhaps the most dangerous situation is where a foal, having been kept completely free of exposure to redworm infection for several months – for example by keeping it off pasture – is then exposed to massive doses of redworm larvae by, for instance, being placed on a heavily parasitised pasture. The foal, having had no previous experience of the parasite, may succumb to the overwhelming infection, suffering from diarrhoea; heavy infestation may also, depending on the particular species of redworm, cause the gut to rupture due to the damage caused to the intestinal wall and blood vessels.

White Worms White worms *(Parascaris equorum)* are found in horses under three years old, with rare exceptions in older horses. The adult worms live in the small intestine and lay eggs in which larvae develop. There is no free-living larval phase, and the only means of infection is by

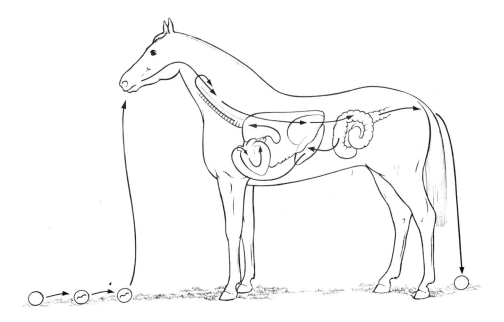

Fig 36 The life cycle of Parascaris equorum *is as follows: the eggs containing larvae are swallowed and hatch in the intestines from whence they migrate through the liver, reaching the heart and passing in the blood stream to the lungs where they eventually reach the air passages and travel to the throat. Here they are swallowed and develop into adults in the large intestines where they lay eggs which are passed to the exterior in the horse's faeces*

a horse eating the eggs. After being swallowed the larvae hatch and escape into the intestines. From here they migrate into the liver, through which they pass in the blood stream, through the heart to the lungs (fig 36). In the lungs they migrate up the airways to the pharynx and are swallowed, returning again to the intestine, where they mature within about eight weeks of having first entered the host. The adult worms may cause intestinal disturbances, and occasionally rupture the intestines in young foals. During their passage through the liver and lungs they cause damage leading to loss of condition and pneumonia. Symptoms include unthriftiness and bronchitis. There are a number of drugs of choice which can be used to treat horses and these should be administered regularly – at one- to two-month intervals – to foals, yearlings and two-year-olds.

PART IV

RECENT DEVELOPMENTS

13

FEEDING AND BREEDING

Feeding horses represents a somewhat uneasy relationship between art and science. During the course of evolution, the horse developed teeth and a gut suited to grasses of varying fibrous content. This system of flat-tabled teeth and hind gut (colon and caecum) with bacterial fermentation of fibre was best suited to the dietary habits of the wild ancestral horse, and had to adapt when man domesticated the species and came to control its diet.

In essence, we cannot change the system of digestion. However, we can use information we gain from research to harmonise diet with the endpoint of digestion, namely growth and health. For years this harmony was based upon the art of feeding with very little scientific information as to the optimal dietary content or quantity. Hay and oats were considered the staple diet, but bran, maize, beans and other legumes were often included in the feed.

Over the years, certain principles of feeding have been recognised in scientific terms, especially where deficiencies or excesses were associ-ated with, for example, colic; and, more significantly, with maldevelopment, particularly of bone. From practical observations and research experience it was found that an imbalance of calcium and phos-phorus caused physitis or other abnormal conditions of bone; and copper deficiency resulted in damaged joints.

Other errors of dietary control known to have adverse effects include over-feeding, especially diets of high energy content to young stock and pregnant mares. The effect of bolus feeding bulk contribution at specific times has also been considered. Horses are naturally continuous feeders, ie they take small amounts over long periods – as opposed to carnivores, for example, whose digestive system is designed to accept an occasional large meal. 'Little and often' is therefore the best approach for horses, albeit within the practical limits imposed by managerial resources.

The wild horse grazes succulent plants containing relatively large amounts of water, soluble proteins, lipids and sugars, but little starch. It eats for short periods throughout the day and night. In contrast, during the process of domestication over several thousand years, man has modified the lifestyle of horses to suit human purposes. The husbandry adopted has been developed by the trial and error of past and present generations.

Diet-related problems have arisen partly because of this restriction on feeding time and partly because of the introduction of 'unnatural' materials, such as starchy cereals, protein concentrates and dried forages. This is not to suggest that we should go back to free-ranging natural habitat-orientated management of breeding stock. This would not be practical, even if it produced satisfactory results with regard to fertility and acceptable growth rates in young stock; the acreage is not available, nor the manpower necessary to herd mares and foals under these conditions.

This book describes the science behind the art of breeding horses; diet is an obvious component of this enterprise. There are many excellent texts on the science of feeding: two examples are given at the end of this chapter. Here, however, the principles considered most relevant to the feeding of breeding stock, rather than the details, are discussed.

GENERAL

Energy is the central function of diet; energy to walk, canter and gallop, to get up and down, to move around the loose box and, importantly, to generate sufficient heat to maintain body temperature. The horse obtains its energy requirements from food. This is digested in its passage through the small intestines but the process is completed in the hind gut by bacterial fermentation. Young foals, up to about the age of two months, are the exception, as a substantial amount of digestion occurs in the intestines because of the particular diet, ie milk.

Digestion is the process of breaking down complex material into simple carbohydrates, fats and amino acids. These are absorbed through the small and large gut into the blood stream. The amount of energy supplied through this process depends in part upon the quality of the feed. This is expressed as digestible energy (DE).

In general, high quality feeds are preferable and their cost may be offset largely by the need to feed less. This is because they contain more digestible energy than the equivalent amount of lower quality feed. This is particularly true of hay which is an integral part of a diet and supplies the fibre necessary for the process of digestion. Further, low quality hay often carries a higher risk of challenge by fungal spores and dust that does so much damage to the lungs of stabled horses.

One of the most obvious difficulties in formulating a diet for horses based on scientific principles is the variation in the content of oats, hay and pasture. The constituents of grass will vary according to the month, or even week, of the year; and hay will differ between batches. Therefore it is difficult to calculate the amount of calcium or protein, for example, that an individual receives daily .

These variations are not particularly important providing general principles are followed. The most obvious risk in young stock is the feeding of high protein diets at a time when the protein content of the grass is high. Excessive protein intake, from time to time, may not be of great consequence to an adult; but in a foal or yearling it may cause an abnormal growth spurt resulting in epiphysitis or contracted limbs (see page 324). Bolus feeding of concentrates may also interfere with the normal development of bones, joints and ligaments.

Another consequence of management relates to the practice of stabling horses and allowing them only restricted exercise. Energy is required to support activity and restriction of exercise, coupled with overfeeding, may have deleterious consequences, especially upon young stock. The horse's body is designed for slow travel over long distances throughout twenty-four hours of every day. Long periods of rest, interspersed with quite severe bouts of exercise, may have quite a profound effect on skeletal development and growth.

The art of feeding is to relate intake, in terms of quality, composition and quantity, to the energy output of the individual. The breeder should consider the effect of diet on 1) sexual performance, ie the need for promoting oestrous cycles and obtaining conception; 2) maintenance of pregnancy; 3) birthweight, maturity and confirmation of a foal; 4) growth and development of the foal. Also 5), when establishing a diet to be given after weaning, the breed and future career of the individual must be considered. Objectives will differ between a Thoroughbred which is to be sold at a yearling sale, one that will be put in training, and one that will not work for a further year or two. Each of these endpoints may be reflected in the diet, although all growing animals should be fed with the aim of allowing the individual to reach its optimal growth and development, without being forced or retarded to meet some artificial endpoint. For stallions and brood mares diet should be aimed at maintaining fitness and health rather than purely show condition.

The science of feeding aims to achieve the endpoints or objectives in a way which takes into account the basic requirements for a healthy animal. A plethora of information regarding various aspects of feeding has appeared in the literature and been taken up by the proponents of feed additives, on the one hand, and owners on the other. Perhaps the best approach for a book devoted to the subject of breeding is to provide the reader with guidelines upon which he or she may build by further reading and/or personal experience.

FEED QUANTITY AND QUALITY

Hay is the central part of a horse's diet for all but the youngest foal. This is because the hind gut depends upon fibre to maintain its normal function. Grass can be used to replace hay, but most breeding stock are maintained on a mixture of the two. Both of these forages vary considerably in quality, and in dry matter content, depending on the season of the year and the climate in which they are produced. Both are suited to the continuous feeding habit of the horse, being consumed slowly over several hours daily, which is preferable to the devouring of concentrates over a short period.

For breeding stock of any age and condition it is a good policy to have hay or grass available for consumption *ad lib*. On average a 500kg mature horse will consume about 10kg of hay in twenty-four hours, although there is a quite wide variation according to individual appetite. The National Research Council (NRC) recommends that, from the last 90 days of gestation and during lactation, a mare should receive about 6kg; a stallion 5kg; a weanling 2.2kg; a yearling 3.7kg of hay containing 8.2 MJ,DE/kg and 88 per cent dry matter. These are only guidelines by which to assess the amount of hay that any particular horse should receive.

The amount of concentrate also has to be taken into account in conjunction with the hay. The NRC defines 'concentrate' as a mixture containing 11.4 MJ,DE/kg and 88 per cent dry matter. According to the NRC, a mare weighing 500kg should receive 2.6kg, a lactating mare 6.4kg, a stallion 5kg during the breeding season and 2.5kg outside the breeding season, a weanling 4.3kg and a yearling 3.6 kg per day.

The term 'concentrates' covers a wide variety of foods from crushed oats to proprietary cubes or mixtures containing controlled levels of protein and other constituents. The protein content of diet is one of the most variable and, perhaps, challenging aspects with regard to the health and well-being of the foetus, foal and yearling.

Protein

Protein is a substance composed of smaller units known as amino acids. Horses obtain these from plant protein. Some, but not all, amino acids are a dietary essential. The quality of protein in the diet depends largely, therefore, on whether or not it supplies the essential amino acids in the proportions required by the body to build tissue. Essential amino acids, such as lysine, histidine, methionine and tryptophan are necessary for the utilisation of other amino acids. Deficiencies in essential amino acids may, therefore, result in poor use of other, non-essential amino acids.

Tissue proteins are continuously broken down into amino acids and re-synthesised throughout the life of the individual. However, because of wear and tear of the tissue (such as sloughing of surface tissue) there is a continual need for dietary protein replenishment. However, these amounts are small relative to the protein required for growth or milk production, and the needs of pregnant and lactating mares, foals and yearlings are particularly high.

The percentage of protein in the diet should be kept within fairly narrow limits for each class of breeding stock. Too much protein carries the risk of triggering growth spurts, particularly of bone, that may result in disproportionate growth and consequent skeletal problems (eg knuckling over at the fetlock, epiphysitis). In older animals, too much protein may affect digestion and lead to the production of toxins in the gut and such conditions as colitis x, diarrhoea and laminitis.

It is usual to describe the upper limit of protein content in the diet as percentage of total dry matter. This should be approximately 12 per cent for a barren mare, and 14 per cent for a pregnant mare rising to 16 per cent in the second half of pregnancy and throughout lactation. A young foal should receive about 18 per cent, a weanling 16 per cent and a yearling about 16 per cent. These upper limits are only approximate guidelines because of the variation that inevitably occurs according to the type of pasture, the climate and the time of year.

Lower limits of dietary protein content are not defined clearly but the general principle is to provide feeds containing the whole range of essential amino acids in a highly digestible form. Good quality hay and spring and summer pasture are the two most reliable sources. In their absence, concentrates can be substituted, such as oats supplemented by appropriate proprietary mixtures containing protein from a variety of sources such as beans, fish meal or milk.

Carbohydrate

Horses obtain the carbohydrates they need from digestion of cellulose and similar plant substances in the large gut. The caecum and large colon grow faster than the small intestine from about the age of three months onwards. This development corresponds to the change in diet from milk to fibrous foods as the foal matures.

Over 80 per cent of a horse's diet is comprised of carbohydrate in one form or another. Digestion (ie direct action by enzymes secreted into the gut) and fermentation (ie breakdown by bacterial action) result mainly in glucose and various short-chain fatty acids which are absorbed into the blood stream through the gut wall. Glucose and one of these acids are converted to a form of carbohydrate known as glycogen that is stored in the liver; whereas others, such as acetate and butyrate, are laid down in the fat pool or oxidised directly. A substantial proportion of the carbohydrate intake, however, is used directly as energy for muscular activity and for the function of organ and tissue systems.

Excess glucose is converted to depot fat, and the amount of glucose present in the blood stream is controlled by insulin. When blood glucose rises, insulin is secreted from the pancreas and by promoting the conversion of glucose to fat and glycogen its action is to reduce too high a glucose level in the blood.

All components of a horse's diet contain substantial proportions of carbohydrate but grain, such as oats, constitutes the major readily available source. It is in this regard that the bolus feeding of grain to young growing stock may have deleterious effects because of the peaks in blood glucose that this type of feeding may produce.

Fat

The diet of horses does not normally contain high proportions of fat; the horse depends on its fat reserves for conversion from carbohydrate in the gut. However, modern diets may be supplemented with dietary fat, often in the form of corn oil. This supplementation is usually confined to performance horses with abnormally high energy requirements rather than breeding stock.

Milk

The young foal, up to the age of about five months, obtains a substantial proportion of its nutrient requirements from milk which provides a significant source of calcium. The composition of milk changes considerably during the twenty-four hours following parturition; but thereafter it remains fairly constant. Gross energy remains at about 260kJ/100g of milk and total solids about 10 per cent on a weight per volume basis.

After their first twenty-four hours, foals require about 150ml of milk per kg bodyweight daily. A Thoroughbred foal weighing 50kg, therefore, receives about 7.5*l* of milk per day.

Calcium and Phosphorus

Calcium (Ca) and phosphorus (P) play interdependent roles as the main elements providing skeletal strength and rigidity. Bone has a calcium:phosphorus ratio of 2:1, although the ratio is narrower (1.7: 1) in the whole body because phosphorus is distributed in the soft tissues in greater proportions to calcium.

Bones continuously take in and excrete calcium and phosphorus, although the balance remains constant. This action facilitates the growth and remodelling of the skeleton as it matures. In effect, this information enables us to aim at a theoretically ideal ratio between calcium and phosphorus of 2:1 in the diet. Unfortunately, there are a number of variables that interfere with this simplistic approach and the reader should be aware of these as follows:

a) The calcium content of the diet may not all be available to the body. It varies between the nature of the feed but, in general, the net availability of calcium lies between 45 and 70 per cent of feed content. However, if significant amounts of oxylates are present the absorption is diminished. A similar effect occurs when inorganic phosphate in the diet is increased.
b) A daily calcium intake of 5g per 100kg bodyweight is required for maintenance – this is about 2.5g/kg of feed. The amount of phosphate required is 2g per 100kg bodyweight or 1g per kg of feed. Excessive calcium intake should be avoided, because it may depress the absorption of other minerals such as magnesium, manganese and iron.
c) The maintenance requirement for calcium and phosphorus is to balance the loss in the faeces and urine, and in the younger animal, for growth. In the pregnant mare, calcium is required for the well-being of the foetus, and in the lactating mare, daily dietary calcium and phosphorus requirements should be increased to about 10g of calcium and 5g of phosphorus, to make good the losses in milk secreted.

Limestone flour, dicalcium phosphate and bone meal are reliable sources of calcium, and 10g of calcium is provided by 28g of limestone or 40g of dicalcium phosphate, which also supplies the requirements of phosphorus.

Copper

Copper deficiency is not a problem in horses as it is in cattle. Nevertheless, there is a relationship between copper intake and a degenerative disease of joints in young horses, a condition known as *osteochondrosis dissecans* (OCD). In this condition, the surface of the joint becomes eroded and cartilage flakes off the surface leaving a roughened, weakened area below. Copper deficiency is thought to affect the enzymes that control the development and health of cartilage.

A dietary intake of 15–20mg/kg of feed is recommended. The extent to which pasture plants extract copper and other trace elements from the soil depends on various conditions such as pH, moisture content and plant species.

Zinc

Zinc is an essential trace element, but grazing pastures affected by industrial pollution may develop signs of toxicity. Dietary intake exceeding 1,000mg/kg of feed may result in painful swelling around the epiphysis, stiffness and lameness.

Manganese

Manganese deficiency is sometimes thought to be related to conditions of knuckling over at the fetlock joint and it is suggested that pasture should not contain less than 20mg per kg dry matter. Manganese may affect enzyme systems involved in cartilage and, therefore, bone growth.

Iron

A dietary concentration of 50mg/kg bodyweight is adequate for growing foals. Iron deficiency does not normally occur in foals and supplementation in the first week after birth should be avoided because it may cause toxic damage to the liver or other organs.

Iodine

Iodine is a component of thyroxine and a deficiency may affect the thyroid gland, causing problems in pregnant mares and foals. However, diets supplemented with 0.1–0.2mg of iodine/kg bodyweight are adequate to meet the needs of horses. Excessive feeding of iodine, such as might occur when large amounts of seaweed are consumed, may cause the foal to develop a large thyroid gland and this can be associated with weakness and, in some cases, death.

Selenium

Selenium is closely involved in the vitamin E complex and the metabolism of certain fatty acids. Deficiency causes a condition known as pale muscle disease (muscular dystrophy), but this is uncommon except in areas where selenium deficiency of the soil exists. Horses require about 0.15mg of selenium per kg of feed for their dietary requirements; higher levels may cause toxicity.

Magnesium

Magnesium is an essential element of inter- and intra-cellular fluids. It is absorbed from the small intestine but a proportion is secreted into the intestinal tract or excreted in urine. To allow for this, the diet should contain about 13mg/kg bodyweight (2g/kg diet) daily. Magnesium deficiency can result in hypomagnesaemia, manifested by nervousness, sweating, muscular tremors, rapid breathing or convulsions.

Potassium and Sodium

Adult horses require about 5g of potassium per kg of feed whereas foals require about 10g per kg of feed. Concentrates are usually rather poor sources but hay contains about 20g potassium per kg and horses fed hay should therefore receive adequate amounts.

Most hay and pasture are rich sources of sodium. The diet should contain about 10g of common salt per kg feed to meet normal sodium requirements.

Foals suffering diarrhoea may become deficient in potassium and, occasionally, sodium.

Vitamins A, D (Calciferol) and E

An adult horse weighing 500kg should receive a daily intake of 50,000iu of vitamin A. Excessive amounts may be toxic. Grazing horses derive their vitamin A from the carotene present in grass. Beta carotene in fresh leafy herbage provides the equivalent of 150,000iu of vitamin A per kg dry matter. Horses are said to be relatively inefficient in converting this beta carotene to vitamin A, but most proprietary feed additives contain this vitamin. Vitamin D enhances the absorption of calcium from the gut and the absorption and deposition of calcium in bone. Daily requirements are about 1,000iu per kg feed. Vitamin D should not be fed in excess. With regard to vitamin E, rations should contain 75iu of vitamin E per kg of feed and slightly more than this in very young foals.

Biotin

Biotin in maize, yeast, soya bean and grass is more readily available than that contained in grain such as wheat and barley. Supplementation is recommended at the rate of about 3mg daily for a 500kg horse to improve the condition of hooves. However, this may take several months.

Water

An adult's body is composed of 65 to 75 per cent water and a foal's is 75 to 80 per cent water. Horses require water as a fluid medium to facilitate digestion and the propulsion of food along the gut, to produce milk, and to replace losses occurring through the lungs, skin, faeces and urine.

Requirements for water vary with environmental temperature and the activity of the individual; they increase during pregnancy and lactation. Horses obtain water from three sources: drinking, natural herbage, and concentrates.

As a rough guide, horses need to consume about 5*l* per 100kg body-weight per day, although this varies greatly between individuals. Water should be made available to all horses at all times. Water purity varies and some sources may have deleterious effects, especially if the horse becomes allergic to impurities in the water supply. These allergies may be manifested as diarrhoea, colic or loss of condition. Diarrhoea can result in water loss leading to serious dehydration. In many areas, the high calcium content of water may have beneficial effects.

PRACTICAL FEEDING

January to March

During the first three months of the year, pasture is at its least nutritious; there may be frost and snow, and paddocks are used mainly for exercise rather than grazing. Barren and maiden mares may, at this stage, be subjected to artificial lighting to stimulate oestrous cycles (see page 74). A rising plane of nutrition may be helpful to these ends and the feeding of good quality hay and about 1.8kg of concentrates per day should help to prime the mare for breeding efficiency.

Pregnant mares entering the final trimester of pregnancy should receive energy intake 10 to 20 per cent above maintenance levels. At least 10kg of top quality hay should be provided, although *ad lib* availability is preferred. Concentrates are less important and pregnant mares may have a reduced appetite during the final six weeks of gestation. They should be given 4 to 6kg of concentrates daily containing about 15 per cent of protein and this should be divided into two or three feeds. Minerals may be supplemented, if complete concentrate feeds are not used, the most important constituent being calcium.

Mares with young foals at foot have the same feed requirements as late pregnant mares, although the amount of concentrate fed may be increased according to appetite. Attention must be paid to maintaining good alimentary function by the provision of bran mashes, the addition of a small amount of paraffin or corn oil to the diet or a small amount of additional salt. These matters have to be addressed according to the firmness or otherwise of the dung passed by the individual. Mares with foals at foot, as with all other categories of mares, benefit by some exercise. If conditions underfoot do not permit this, exercise in a yard or even lunging is beneficial. This exercise also helps the uterus to cleanse itself of debris. This latter aspect is particularly important in the first two or three weeks after foaling.

There is generally no need to supply extra feeding for the foal, apart from allowing it ready access to the hay and feed supplied to its dam.

Yearlings require the best quality feed and frequent small feeds of concentrates. Protein should be restricted to 16 per cent by weight of food intake. Sanfoin (*alfalfa*) hay provides protein, calcium and fibre requirements, and if available, may be used instead of concentrates.

April to June

This is the period of maximum protein content in grass, and pasture feeding is therefore a prime consideration. Mares permanently at pasture at this time require little, if any, additional feeding. The flush of grass will promote breeding activity in barren and maiden mares not already undergoing oestrous cycles. Pregnant mares will receive a boost in dietary content commensurate with their increasing demands for the late stages of foetal development. Lactating mares gain a similar boost toward the production of milk, and the young foal will enjoy adequate energy and protein to ensure that a rapid rate of growth occurs at this age period.

If mares and foals are stabled for at least part of the day, access to good quality hay should be provided. Mares recently mated (maidens, barren mares and mares with foals at foot) require adequate feeding levels to ensure that their foetus is not lost due to dietary deficiency. It is postulated that twins can be reduced to singletons by 'starving' mares. If this is in fact the case, and many dispute its efficacy, it may also be true that starvation can cause the loss of a single foetus; much of the debate centres around the term 'starvation'. Most proponents of the effectiveness of starving mares with twins in the early stages of pregnancy do not define the energy intake. Nevertheless, once conception has occurred, no pregnant mare at any stage of gestation should be without a reasonable level of nutrient energy.

Foals and yearlings on spring and early summer pasture require little, if any, diet supplementation. If sufficient acreage is available they are best left out day and night. As the pasture growth decreases according to the season, supplementation with concentrates and good quality hay may be started. The objective during the foal and yearling stage should be to ensure a steady level of energy intake avoiding, as far as possible, the promotion of growth spurts and lag phases as a result of marked swings in dietary content and quality. Ensuring adequate levels of copper and a 2:1 calcium:phosphorus ratio is equally important.

July to September

Feeding regimes should be similar to those adopted for the previous quarter, taking into account the reduction in the nutrient content of grass. It is a period when pregnant mares are carrying embryos which are developing with a high degree of cellular differentiation, rather than a substantial increase in size. Nutritional requirements therefore should be maintained on a continuing basis. In the case of lactating mares, the high energy requirements cease abruptly at weaning. There should be a corresponding reduction in dietary intake to remove the nutritional drive to milk production. However, it is unwise to reduce nutrient intake in a pregnant mare to below maintenance levels.

During these months, yearlings may be prepared for sale. According to managerial policy, they may be brought in from the paddock for this purpose, their nutrient and energy intake increased dramatically and their exercise limited. This may produce the desired effect of laying down fat and therefore aiding their appearance for sale, but it may also put the individual at risk to over-reaction from slight trauma and minor abrasions. Joints and limbs may 'fill' due to 'humour'. Firm, fibrous or bony reactions can occur under these circumstances, thereby producing blemishes which, although often of no importance, may deter prospective purchasers.

October to December

Pasture growth may occur in autumn, depending upon prevailing climatic conditions. Protein content may increase but not usually to a degree that need cause concern. Pregnant mares are beginning to enter the stage of maximum demand for foetal growth during these months and rations therefore should be increased. Mares turned out day and night should have access to hay and some concentrates, particularly mares due to foal in the first quarter of the following year.

Weaned foals require access to good quality hay *ad lib* and concentrates containing about 15 per cent protein. Calcium and copper supplementation may be necessary depending on the materials chosen for the diet.

SUGGESTED FURTHER READING

Hintz, H.F. *Horse Nutrition:A Practical Guide* (Arco, New York, 1983)
National Research Council (US) *Nutrient Requirements of Horses* Sub-committee on horse nutrition (National Academy Press, Washington DC, 5th edition 1989)
Frape, D.L. *Equine Nutrition and Feeding* (Longman Scientific and Technical, Essex, 1986, revised reprint 1990)

14

ULTRASOUND SCANNING: SEEING IS BELIEVING

Ultrasound (echography) has made a significant contribution to the diagnostic capability of the veterinary and medical professions. In equine medicine, the technique is now used in the evaluation of diseases of internal organs such as the liver, lungs and kidneys and in the assessment of tendon injuries. However, its first application and general use was among pregnancy diagnosis and interpretation of ovarian activity in mares.

Echography is now used routinely for pre-service examinations of mares, pregnancy diagnosis and in the assessment of foetal health. It can also be used on the external genitalia of the stallion to provide accurate information regarding size and consistency of testes. It is also of use in identifying the presence of oedema or fluid in the uterus and the presence of cysts and other abnormal structures in this organ.

This chapter describes the physical principles on which echography is based, the techniques used for medical appraisal of the genital organs of mare and stallion, and the benefits of diagnosis currently available for breeding purposes.

PHYSICAL PRINCIPLES

Echography means literally the study of echoes. The technique is based on sound waves of an extremely (ultra) short wavelength, hence the alternative terminology of ultrasound and ultrasonography. Colloquially, the process may be described as scanning.

Applied to the reproductive system, it enables us to see where once we could only feel. For example, ovarian palpation described in chapter 3 depends entirely on the operator's interpretation of the size and consistency of the ovaries held between thumb and fingers. The presence of a follicle is, by this means, determined by comparing a fluid sac with the hard or firm feeling of the substance of the stroma of the ovary. Size may be estimated according to the size of the fingers, eg if the fluctuation extends across the tips of two fingers, the follicle may be about 3cm

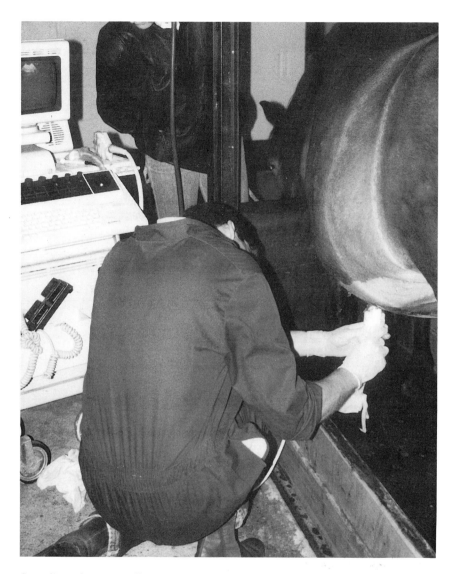

Sampling of amniotic fluid (amniocentesis) can be performed during pregnancy by means of a needle inserted through the abdomen, guided by ultrasound

(overleaf) *Fig 37 Audible sound consists of a disturbance in the molecules in the air due to waves of alternate compression and rarefaction of the molecules as illustrated. Ultrasound works on the same principles but the source of energy is alternating electrical pulsations produced by special crystals in the scanner probe (see fig 38)*

AUDITORY   SYSTEM

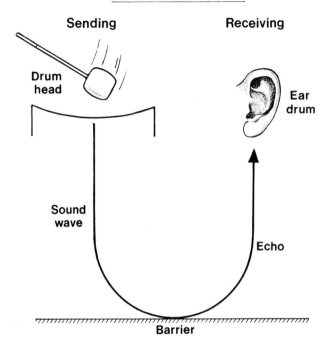

ULTRASOUND   SYSTEM

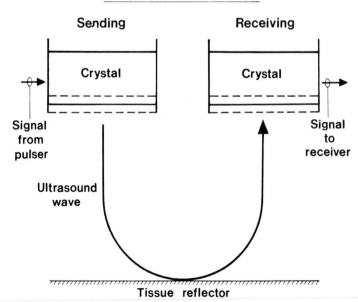

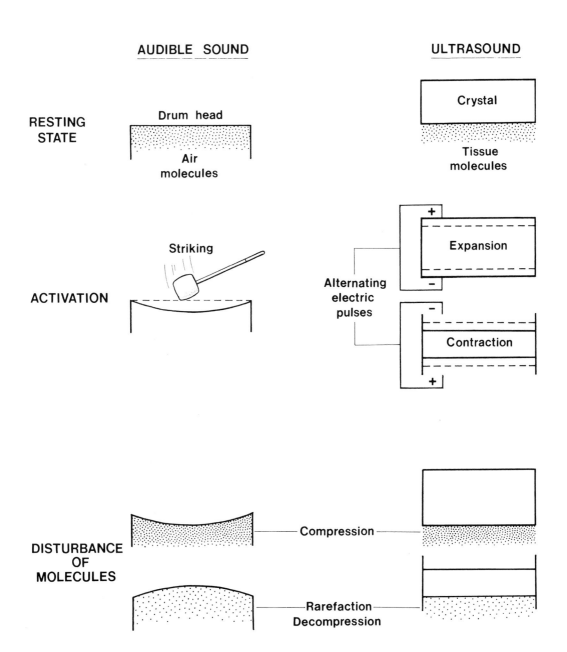

AUDIBLE SOUND

ULTRASOUND

RESTING STATE

Drum head

Air molecules

Crystal

Tissue molecules

ACTIVATION

Striking

Alternating electric pulses

+
Expansion
−

−
Contraction
+

DISTURBANCE OF MOLECULES

Compression

Rarefaction Decompression

diameter. Ultrasound enables us to see, photograph and measure accurately the contents of the ovary.

Another use of the technique is in pregnancy diagnosis. At 42 days from last mating the swelling in the uterus (see page 155) can often be felt per rectum with reasonable ease. However, in some mares this is more difficult and there is a certain element of inaccuracy relating to the experience of the operator. If ultrasound is used, the presence of a foetus can be distinguished clearly but, even more important, it provides an accurate diagnosis of pregnancy as early as 12 days following conception. It can also identify twin conceptions, as will be described later in this chapter.

PHYSICAL BASIS OF ULTRASOUND

Audible sound travels through air at about 330 metres per second. Wavelength is the distance between the alternating wave of vibrations or pulses occurring in disturbed molecules of air brought about by the source of sound. The wavelength of audible sound ranges between 2 and 2,000cm. The frequency at which the vibrations occur is measured in hertz (Hz) units. One hertz is one cycle per second and a megahertz (MHz) is one million cycles per second. Diagnostic ultrasound has a wavelength of less than 1mm and a frequency of 1–10 MHz. At this 'ultra' level, sound cannot pass through air but it travels through tissue at an average of approximately 1,540 metres per second, although the speed varies according to the tissue.

Ultrasound scanning consists of a system by which ultrasound waves are emitted from crystals housed in a probe. The waves travel in a straight line until they are either absorbed, as they become too weak with distance, or reflected (echoed) back as they meet a surface; just as audible sound

(previous page) Fig 38 Illustrating audible and ultrasound (inaudible) sound waves. A drum (top) sending out sound waves which are reflected by a barrier (eg, a mountain) and returned to the ear drum where they set up vibrations that are perceived by the brain as sound. In the ultrasound system short waves are sent out from a crystal, reflected by tissue and the echo received by the crystal where it is electronically converted into a picture (echogram)

Ultrasound waves travel through the body tissue until they reach the border of the tissue when some are reflected; others pass on to be reflected at the next barrier, hence the building up of a two dimensional picture on the ultrasound screen

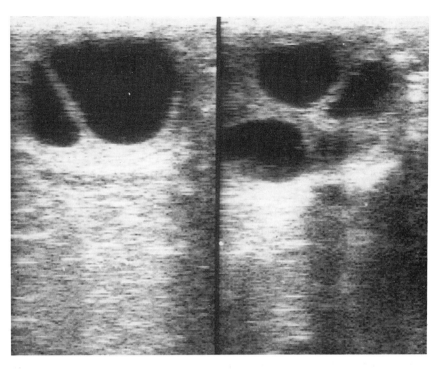

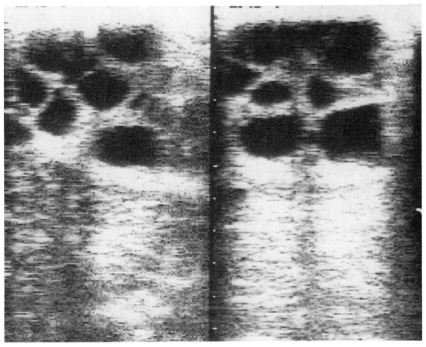

waves may be echoed back from a mountainside. The reflected echoes are received back by the crystal and interpreted into a visual form by an electronic process. The ultrasound transducer or probe contains a large number of crystals thereby enabling a two-dimensional picture to be established over a broad front.

In summary, ultrasound scanning consists of sound waves dispatched from a source and received back in a form dependent on the distance and material through which they pass. At each surface some or all of the waves are reflected and the picture obtained can be interpreted in terms of distance from the source thereby allowing size and shape to be determined. The size and shape can be viewed only in a two-dimensional cross section, but if the probe is moved over the area some three-dimensional concept can be achieved, hence the term scanning.

There are a number of variations in the terminology of ultrasound which need not concern the reader in detail. However, the following terms may be used commonly by veterinarians:

3, 5 and 7 MHz probes

These are sources of ultrasound at different frequencies. The higher frequencies (vibrations) have less power of penetration but a higher resolution (definition) in the picture that results. The 7 MHz is used therefore for scanning such structures as tendons where penetration of no more than 3 or 4cm is all that is required but a high resolution is necessary for interpretation of injury. In reproductive work, the 7 MHz probe is useful for examining the placenta where a great deal of detail is required. The probe used most often is a 5 MHz, which has a penetration of about 10cm and good resolution for identifying follicles and early pregnancies. The 3 MHz probe has a penetration of 20–25cm and although less detail is provided, it should be possible to make some interpretation of structures at a considerable depth within the body. They are used, therefore, for scanning a pregnant uterus in the latter stages of gestation and for viewing internal organs such as the kidney, liver, etc.

Ultrasound scan of left and right ovaries. There are three 2cm diameter follicles in the right ovary, and a 3cm diameter follicle in the left ovary, with smaller follicles immediately adjacent

Ultrasound scan of left and right ovaries with numerous immature (non-ripe) follicles present

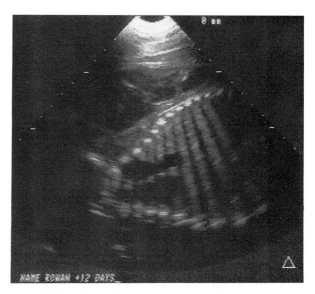

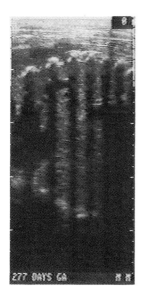

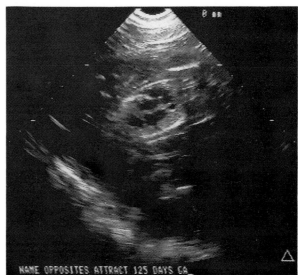

Linear scan taken through the abdomen (transabdominal) of a 273-day foetus, showing ribcage and heart (shadow on extreme left)

Transabdominal sector scan showing foetal heart and aorta to left

Transabdominal sector scan showing foetal heart with four chambers: left and right ventricles and atria

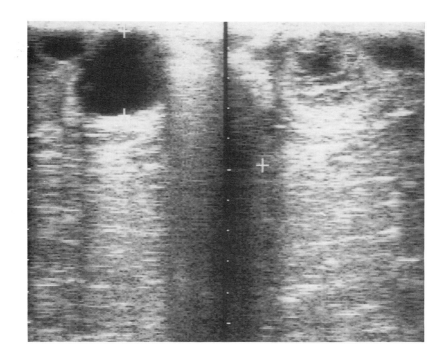

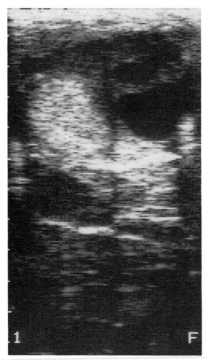

Scans of ovaries showing yellow bodies (arrowed) of approximately 5 days of age

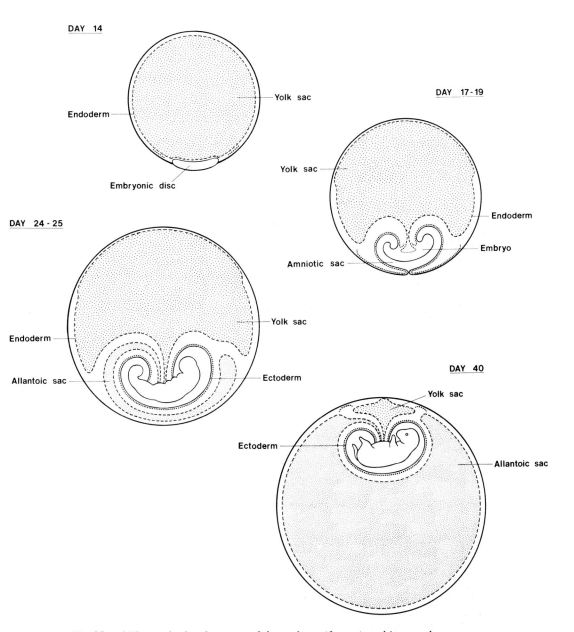

Fig 39a–d The early development of the embryo (foetus) and its membranes between Days 14 and 40 post-conception. At Day 14 the cross-sectional area is about 0.5sq cm, and this increases around forty-six times to approximately 23sq cm at Day 40. Note that the yolk sac diminishes in size as the allantois increases and combines with the chorion to form a true placenta

Linear and sector probes

These terms refer to the arrangement of crystals providing the source of ultrasound. In the linear probe, as the term indicates, they are arranged in a straight line and the effect is a picture in the shape of a rectangle. With a sector probe, the source is concentrated and spreads out in a wedge-shaped view. The linear probe is ideal for situations where the length of the transducer can be readily accommodated, eg over the uterus. The sector is used when a view through a small aperture is required as, for example, between the ribs to observe the inside of the chest for the heart.

Routine examinations

Ultrasound is used in conjunction with the veterinary programme described in chapter 3 (page 99). It is used to visualise the ovaries when a clear interpretation of ovarian palpation cannot be established. In particular, it helps to confirm whether or not ovulation has occurred. In many instances the follicle that ovulates may still resemble a follicle on palpation because of its fluctuating nature: instead of the follicle being filled with fluid, it is filled with the blood that escapes at ovulation and fills the cavity. The yellow body is formed from the cells that invade the blood clot. Therefore, the operator can confirm the occurrence of ovulation because the clear fluid in the follicle has been replaced by the speckled appearance of blood.

Follicles may be distinguished readily and measured. Before the use of ultrasound, palpation of of the ovaries often identified a follicle but could not determine the actual number present in the ovary. Sometimes as many as 10–20 follicles are present and these would not be apparent on palpation. The same discrepancy occurs when two follicles are close together. Ultrasound shows these quite clearly.

Ovarian scanning is not, unfortunately, an absolute answer to pre- and post-service examinations of the ovary. A number of aspects are still difficult to interpret for reasons which readers should understand in order to obtain a balanced account, not only of veterinary capabilities, but also of the biology of reproduction.

We may indeed be able to identify follicles and calculate their exact size and shape. But this information, valuable as it is, does not provide a very important part of the jigsaw puzzle of sexual behaviour. Follicles mature and ovulate under the influence of hormones (oestrogen, progesterone, luteinising hormone and follicle-stimulating hormone) whose concentrations are changing in the blood stream from day to day, if not from hour to

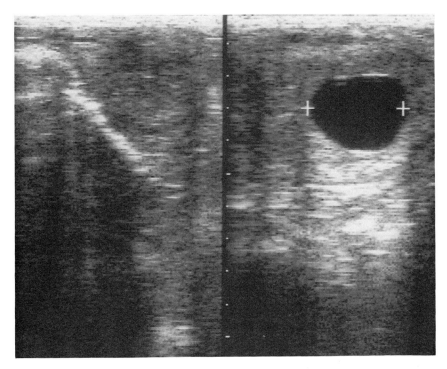

*Scan performed 17 days from last mating, showing single foetus in right uterine
horn (diameter of sac 29mm). Note there is no sac in left horn*

*Scan performed 35 days from last mating, showing single foetus in right uterine
horn (diameter of sac 43mm). Note there is no sac in left horn*

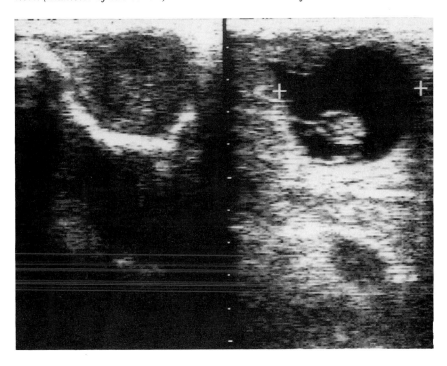

hour. We have no information about these changing patterns at any given time of examination. However, even if we had this information, we should still be lacking knowledge about the receptor status of the target tissue (eg the lining of the follicle). Receptors represent the keyhole where hormones are the keys. If there are no keyholes, the keys have no means of exerting any action. And the number of receptors for LH per follicle is directly responsible for the action that LH can have on the follicle with regard to ovulation. When we feel or view a follicle we have no means of determining the number of receptors in each follicle; and it is not surprising therefore that follicles do not always ovulate according to our interpretation of when they should, based on size, shape and feel. The fact that the majority of our interpretations are correct indicates that, in general, we are able to assemble sufficient evidence on the basis of the various examinations we can undertake within the limits of our competence in these matters.

Once a mare has been mated and ovulation has occurred, it is necessary to establish, at the earliest opportunity, whether or not the mare is pregnant. The background to this need is discussed on page 155.

Ultrasound scanning allows a positive diagnosis to be made as early as 12 days after ovulation (usually about 14 days after mating). At this stage, the developing embryo consists of a fluid sac or vesicle, measuring approximately 10mm in diameter, that appears as a black hole on the scan against the whitish surroundings of the uterine wall.

Also at this stage the walls of the vesicle are composed of two layers of cells, exoderm and endoderm (outside and inside layers). In one small area of the wall, a concentration of cells occurs which is the start of the embryo proper. This area does not appear on the ultrasound picture.

Between about 10 and 16 days post ovulation, the vesicle is highly mobile and travels quite rapidly around the horns and body of the uterus. It has been suggested that this activity prevents the release of prostaglandin from the uterus and thereby protects the yellow body in the ovary that is necessary for maintenance of pregnancy. If the mare is not pregnant, prostaglandin is normally released at about Day 16 but contact of the vesicle with the uterine wall is thought to inhibit its release from that stage onwards.

This mobility of the vesicle has two practical consequences. When scanning prior to the 16th day from ovulation, it is essential to scan every part of the uterus, both horns and body, otherwise it may not be identified. After the vesicle is fixed at Day 16 it will be identified more readily in one of two positions at the base of one or other of the horns.

Mobility is also important in relation to the identification of twins and

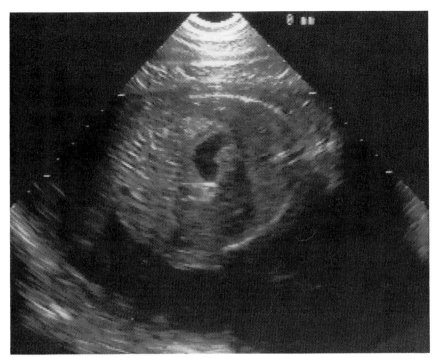

Transabdominal sector scan showing a cross-sectional view of a foetus at 125 days gestation, with foetal stomach at centre surrounded by liver on the left and spleen on the right

Similar view but slightly behind photo 117, showing liver and four blood vessels in cross-section

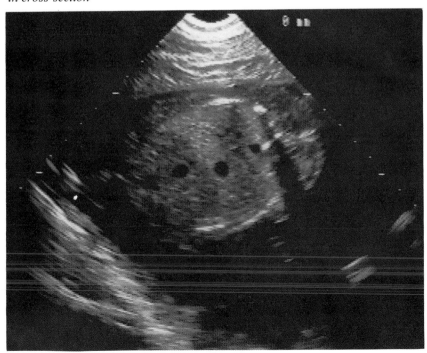

in dealing with them effectively, as discussed below.

As the vesicle increases in size so does the complexity of its structures. And a corresponding change in pattern can be observed on the ultrasound screen.

The landmarks of scanning are shape and size of the vesicle (diameter 4cm Day 16–23cm Day 40), the appearance of the foetus itself by Day 22 and its heartbeat by Day 24. From this stage onwards the line dividing the diminishing yoke sac with the expanding placental membrane (all-antois) is clearly visible. By Day 40 the yolk sac has shrivelled and the placenta surrounds the foetal body to which it is attached by its cord.

From Day 40 onwards the foetus and its membranes become increasingly distinguishable into the components we recognise at birth, namely amnion, amniotic fluid, allantoic fluid, placenta and cord. The volumes of fluid increase rapidly and the uterus expands to accommodate this increment.

At this stage, the penetration of a 7 MHz probe starts to become too short for an overall view of the foetus. This restriction is apparent by Day 60 and by Day 80 it is necessary to view the uterine contents with a 3 or 5 MHz probe against the abdomen. This approach may be used throughout the rest of pregnancy until full term. However, as the foetus grows it cannot be viewed in its entirety, even from an external point of view. Nevertheless, the various parts of the body may be scanned so that we can assess heartbeat, state of liver, lung, kidneys, etc.

It should be emphasised that, in the very early stages of pregnancy, differences in size of the vesicle and appearances of various structures and heart beat may vary considerably from day to day. Calculations made from the last date of service, therefore, should take into account the fact that if ovulation occurred three days later the ultrasound scan would be of a 14- rather than a 17-day pregnancy. The descriptions used here are based on time from ovulation rather than from the last date of service. Once the gestational age is above 25 days, the discrepancy becomes increasingly less significant.

Ultrasound scanning also enables us to diagnose whether or not the foetal foal is presented with its head towards the cervix. A very small percentage of pregnancies terminate with the foal presented backwards. Experience with scanning shows that the foetus is naturally positioned with its head towards the cervix by 180 days of pregnancy, although many will achieve this position prior to 180 days.

The activity of the foetus, clarity of the surrounding fluids and health status of the placental membrane may be assessed with ultrasound and

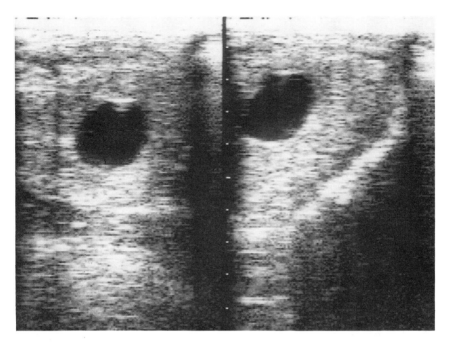

Twins present, one in each uterine horn, scanned at 16 days from last mating

Scan of the uterus of a mare 17 days from last mating, with twins lying adjacent to one another in the same horn

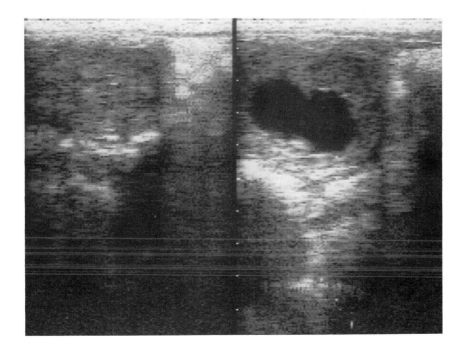

these will provide an indication of health or ill health of the foetus. Sluggish activity with numerous particles in the fluids may be signs of ill health.

The foetal heart rate decreases with gestational age, being in the region of 200 beats per minute in the first third and below 100 beats per minute in the last third of pregnancy. Heart rate can also be used as a guide to foetal health or ill health.

Another form of ultrasound, namely the Doppler technique, can be used to study blood flow in the heart and aorta. This technique is employed routinely in the assessment of infants *in utero* but has not as yet been extensively employed on the foetal foal.

Measurement of foetal components can be used as a guide to foetal growth. Those most generally studied are the width of the aorta and the diameter of the orbit of the eye.

TWINS

Before ultrasound scanning was introduced into routine studfarm examinations, the incidence of twins in Thoroughbreds had remained stubbornly at about 5 per cent. In the 1940s, when rectal palpation was first practised on a wide scale, it had been hoped that twins could be avoided by withholding mating when two follicles were palpated. Various stratagems were tried, such as arranging mating between the ovulation of one follicle and that of the other. Another approach was to flush out the uterus on Day 5 post ovulation in the hope of removing one fertilised egg before the other had entered the uterus. A further approach was to identify the presence of twins by rectal palpation in each horn of the uterus as early as possible, which was usually around Day 30 post mating, and to squeeze one of the two foetal sacs. This approach had some success, but often resulted in either the twin pregnancy continuing or both foetuses being lost.

When the twins occurred in the same horn of the uterus it was not possible to distinguish between the two and squeezing was never attempted in these cases. Overall, the problem of twins remained until ultrasound provided the answer by enabling us to a) identify the presence of twins at a very early stage (Day 12–20); b) establish their exact site and whether they were situated together or one in each uterine horn; and c) determine if efforts to obliterate one member by squeezing had been successful.

Today, mares are usually scanned about 15 days from ovulation. Some clinicians prefer to make their first examination a day or two before this. In either case, the vesicle is mobile and if two are present together there is a chance that they may be gently separated before crushing.

The disadvantage of this is that the mobility of the vesicle may allow it to escape from between the fingers when the operator is attempting to crush it. It may then travel along the horn or into the body of the uterus. After Day 16 the vesicle is usually in a fixed position and can be crushed quite easily. Nonetheless, even at this stage, if the two vesicles are adjacent to one another it may be impossible to crush one without affecting the other. In all cases, the judgement of when and how to squeeze is a personal one related to the experience of each individual.

If twins are situated in separate horns, squeezing has about a 90 per cent success rate. However, this rate decreases markedly after about Day 21 of pregnancy. If twins are adjacent, there is a 50 per cent chance of reduction by natural means as one vesicle may interfere with the nourishment of the other to its disadvantage so that it is gradually eliminated by about Day 35. It may be decided, however, that the risk of twins continuing is too great and the mare is then given prostaglandin to abort both foetal sacs so that she may be re-mated. This decision has to be taken before the development of the endometrial cups at Day 35 or else the mare may not return to fertile oestrus for many months.

The overall effect of routine ultrasound scanning has been to reduce the incidence of twins being aborted or carried to full term to less than 1 per cent. Those that are not eliminated before Day 35 may be due to a number of circumstances: a) a deliberate decision on the part of an owner to allow a diagnosed twin pregnancy to continue in the hope that one of the twins will be eliminated naturally; b) failure for a diagnosis to be made because the mare has not been examined or the examination has failed to identify a twin conception. This is more likely if the examination is performed before 16 days following ovulation when the vesicle is mobile. However, even after this stage mistakes may be made. If an examination is performed at the 40-day stage, such errors are extremely uncommon. Finally, c) the presence of cysts in the uterus may confuse the diagnosis both of singletons or twins because they can give the appearance of an early stage pregnancy. Larger cysts may obscure the presence of a vesicle. In these circumstances it is important to scan as frequently as necessary to identify the presence of a heart beat or, importantly, of two heart beats.

THE NON-PREGNANT UTERUS

The image of the uterus when a mare is in oestrus is quite characteristic. The folds of mucosa may be seen forming a pattern due to their water content (oedema). This is particularly apparent in mares that have foaled recently. In aged mares, and those with an infected uterus, there are often accumulations of fluid in the lumen of the horns or body.

ABNORMAL CONDITIONS

Ultrasound imaging may be used to distinguish granulosa cell tumours from ovarian haemorrhages. Uterine cysts and pyometritis may also be identified thus aiding in the diagnosis of uterine enlargement.

15

ARTIFICIAL INSEMINATION AND EMBRYO TRANSFER

ARTIFICIAL INSEMINATION

Artificial insemination (AI) in the horse is practised worldwide on all breeds (see page 130). There are many advantages from a veterinary, managerial and commercial viewpoint.

Veterinary aspects The main veterinary advantages relate to reducing the spread of infection and containing epidemics while breeding continues.

The spread of infectious diseases, such as herpes virus and *Streptococcus equi* (strangles), is enhanced by the movement of horses between premises and even within the confines of one studfarm. Taking semen to the mare reduces the risk of diseases spreading because it precludes the need to move animals.

The same principles of disease prevention apply between individuals. For example, if a mare has been treated for a *Klebsiella* or *Pseudomonas* infection of her genital tract, but insufficient time has elapsed to be sure of her being free from any infection, AI can be used without any risk to the stallion or to other mares mated subsequently. Susceptible mares (see page 254) may also be mated without risk of their succumbing to the coital challenge inherent in natural service.

A further advantage is that the quality of the semen can be assessed and an evaluation made of the stallion's breeding potential.

Managerial aspects During the course of routine pre-service examinations there are occasions when AI is extremely useful; for example, when a mare has developed a sizeable mature follicle but does not respond to the teaser (page 77) or appears to be in oestrus despite a zero level of blood progesterone. This situation occurs fairly frequently in the early part of the year in maiden and in some barren mares when the follicle ovulates without the mare being mated. This is sometimes referred to as a 'silent ovulation'.

The predictive timing of ovulation to within hours has always been difficult and the advantage of AI is that small quantities of semen can be inseminated daily or twice daily. This is particularly useful if a stallion or mare is subfertile and it is necessary to place semen in the uterus within hours of ovulation.

In an AI programme, semen is collected from the stallion(s) daily, and one or more ejaculates divided so that each mare can be inseminated at least once daily while in oestrus. This eliminates the need to identify the optimal day of mating and prevents spread of infection.

Commercial aspects The main financial benefit of AI is in the increased capacity it allows for each stallion. Other financial advantages include the reduced amount of manpower and veterinary supervision of mating programmes required and the fact that the cost of transporting mares is eliminated.

Semen can be frozen and used to provide numerous offspring from the stallion. It is this capability that has caused Thoroughbred breeders to fear that the market may become swamped with the progeny of one stallion at the expense of others.

Semen collection

Semen is collected by means of an artificial vagina (AV). There is a wide variety of models, and some are illustrated here. The AV consists essentially of a tough outer metal tube, with handles attached to facilitate control by the operator. The case is fitted with a rubber or latex lining. Water is introduced between the case and lining to maintain the correct temperature.

The AV can be lined with a disposable plastic lining to reduce the risk of transferring infection between stallions. The internal temperature of the lining should be in the region of 113–122°F (45–50°C) at the start of the procedure. Precautions must be taken to protect the semen during and after collection because spermatozoa are fragile. Sudden increases or decreases in surrounding temperature must be avoided and all equipment should be cleaned scrupulously.

Successful collection depends upon appropriate temperature and pressure on the penis; flexibility and lubrication of the lining; and sympathetic handling on the part of the operator. Most stallions can be trained to ejaculate into an AV, and even stallions used for natural breeding may tolerate collection on isolated occasions if an oestrous mare is present. Stallions kept entirely for AI are often trained to mount a dummy.

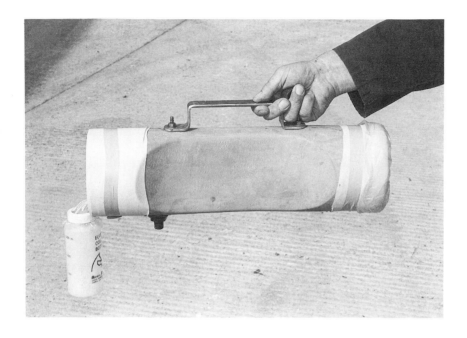

The Cambridge model artificial vagina (above), and assembled with disposable liner and collection model (below)

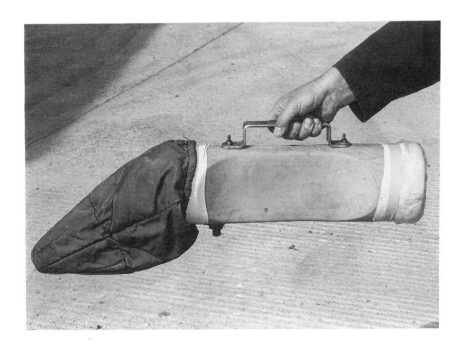

CSU artificial vagina assembled and ready for use, with the protective ruff in place

Semen handling

Semen for AI is used fresh or deep frozen. Fresh semen may be maintained at room temperature for up to an hour or two before being inseminated. It can be mixed with extender fluid which contains protective energy-rich substances to enhance survival of the sperm.

A semen sample may be inseminated in its entirety or split into several samples. It is essential that each sample is handled with maximum care at all times to ensure that the sperm are not damaged.

The number of samples obtained from any given ejaculate depends on the volume and number of live normal sperm present in the ejaculate. In this respect, Thoroughbred stallions may have inferior quality semen compared with other breeds. In practice therefore, their ejaculates are usually divided into two or three samples only, to maintain fertilising capacity.

Fresh semen may be transported to the mare(s) providing insemination takes place within about two hours of collection, although the sooner it is inseminated the better.

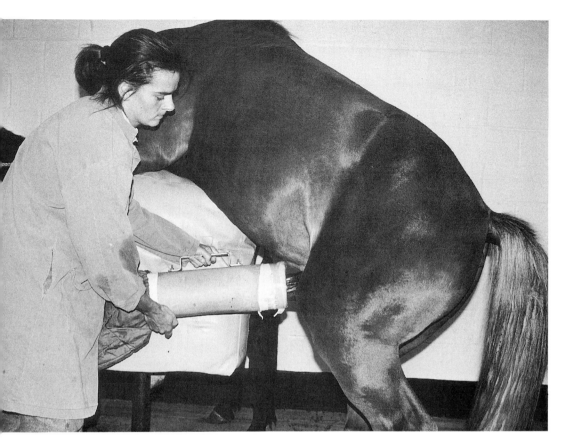

Semen collection from a stallion mounted on a dummy mare

Evaluating semen

Spermatozoa are fragile and it is important not to subject them to changes of temperature. The container into which they are collected should be about 100°F (38°C). Evaluation of a sample's quality is based on the total number of normal motile spermatozoa in the ejaculate. This must be assessed rapidly because sperm tend to lose their motility quickly when maintained in the seminal fluid. It is something of a paradox that the seminal fluid, which is contributed by the accessory sex glands (page 106), is a poorly nourishing vehicle for the sperm and may even be toxic. For this reason also it is necessary to inseminate mares as soon as possible after collection of the semen.

In evaluating the motility of the spermatozoa it is customary to suspend the spermatozoa in a special fluid known as extender. This is also the vehicle used to allow artificial insemination with raw and deep frozen semen.

Motility is measured by placing small samples of the semen in extender fluid at 38°C to achieve a dilution of 1 in 20. The sample vials are maintained in an incubator and drops placed on a clean glass slide so that the sperm can be observed under the microscope. The spermatozoa are assessed individually for progressive forward movement. The result is expressed as a percentage of motile sperm. Further assessments are made every half hour until motility ceases.

Once motility has been evaluated the number of spermatozoa per ml of semen is measured. Special instruments to provide a digital readout of the sperm count are available for this purpose.

Evaluation of the ejaculate is important in deciding the number of samples that can be obtained for insemination. It is generally assumed that AI requires 500 million normal motile spermatozoa per insemination dose. This is only a rough guide because there are a number of variables that affect the pregnancy rate irrespective of the number of sperm inseminated. In practice, up to 80 per cent fertility should result from insemination if up to four oestrous cycles are used, compared with up to 70 per cent in the first oestrus depending on the age and fertility status of the mare. At levels below 500 million sperm per dose, these rates fall quite markedly.

Storage

Storage over prolonged periods has been achieved with great success in all species. Preservation is based on deep freezing and re-activation after weeks, months or even years following collection.

Freezing semen

In the early 1950s it was found that, if stallion spermatozoa were separated from the seminal fluid, re-suspended in glycerol and glucose and frozen at 79°C, about 25 per cent were still motile after thawing. Since then, successful modifications have been made to the process as follows: a) spermatozoa must be mixed with an appropriate vehicle (extender); b) they must be cooled slowly from body temperature and exposed to cryoprotective agents such as glycerol or sugar; c) they must be packaged in appropriate single dose containers, usually referred to as straws, and the slow initial freezing process changed at +5°C to rapid freezing below −100°C and then stored at −196°C; and d) on thawing they must be warmed rapidly to 37°C and used for AI shortly after thawing.

Fertility rates

The percentage of mares that become pregnant after insemination during a period of oestrus ranges from about 50 to 70 per cent when using fresh semen and from about 25 to 55 per cent with deep frozen semen.

The results vary with techniques of collection, transport, storage, extender employed and between individual animals. Natural fragility of spermatozoa differs between stallions.

EMBRYO TRANSFER

The technique of breeding by embryo transfer was first developed about fifty years ago when scientists found that it was possible to extract a fertilised egg from sheep and implant it into another ewe if the sexual cycle of that individual was in a receptive state. The embryo then developed in the recipient, resulting in full term delivery. This discovery enabled breeders of sheep, cattle and other species to produce large numbers of offspring from individual females. The technique has been applied to horses relatively recently. However, like AI, its use in Thoroughbreds is still prohibited although it is an established method of reproduction in other breeds throughout the world. Surprisingly, the UK has lagged behind somewhat in this important area of modern horse breeding, in spite of the fact that a primary centre of research (The Thoroughbred Breeders' Association Equine Fertility Unit) is based in Newmarket under the directorship of Dr W.R. Allen.

The technique offers the following advantages: 1) Individuals of high quality or commercial value can produce more offspring than they would following natural breeding methods; up to eight embryos can be produced by each mare during one breeding season. 2) It eliminates the need for competition and performance horses to be out of action while they carry a pregnancy to full term. 3) It improves the chances of producing foals early in the breeding season. 4) It allows an opportunity to breed from subfertile or aged mares which are capable of producing a fertilised egg but not of carrying a pregnancy to full term.

When the donor and recipient mares have been selected, it is important to synchronise their ovulations. This can be achieved by the administration of prostaglandin and human chorionic gonadotrophin (HcG). The embryo should be transferred between Days 4 and 9 following ovulation of the recipient mare.

Embryo collection can be performed surgically, by gaining access to the uterus via an incision through the flank, or non-surgically. Embryo collection should be performed six to eight days post ovulation; the embryo does not reach the uterine body until Day 6 and, after Day 8, it is less viable.The non-surgical method, currently the most favoured technique, involves insertion of a catheter through the cervix. A special medium is then flushed through this into the uterus. When the medium is drawn off under suction, the embryo floats back through the catheter and into the collection vessel. Embryos cannot always been seen with the naked eye and, therefore, the medium is passed through a very fine filter. The embryo should be washed several times in a solution of phosphate buffered saline containing foetal calf serum. This removes any cell debris from the embryo. After washing, the embryo is examined under a microscope to ensure that it is not damaged and that there is no debris remaining. If the transfer does not take place immediately, embryos may be frozen for subsequent use. Like collection, transfer can be performed surgically, by a flank incision to expose the uterine horn, or non-surgically, releasing it into the uterine body from a pipette inserted through the cervix.

The success of embryo transfer is influenced by many factors, including the reproductive status and health of recipient and donor mares, the expertise and experience of the operator, and the degree to which sterile conditions are provided. The success rate for obtaining embryos ranges from 45 to 60 per cent and the pregnancy rate for recipient mares is in the region of 60 to 70 per cent following non-surgical transfer and 80 per cent following surgical transfer. For both collection and transfer, non-surgical techniques are preferred.

At present, in practical terms, success depends on obtaining a healthy fertilised egg from the uterus of the donor mare. However, research is currently under way to develop techniques for fertilisation of the egg in the laboratory. The first foal to be born following this technique was reported in July 1990 by Dr Eric Palmer and his group from INRA, France. This technique does not have great practical applications but it will become a valuable research tool.

Another technique with exciting potential is that of splitting the egg after removal. Each portion of the embryo is transferred to recipient mares, or returned to the donor mare to develop until full term. Identical twins have been produced using this method and, in the future, it may be possible to obtain more than two live foals from a single embryo.

INDEX